ENCYCLOPAEDIA OF
NANO LABS

ENCYCLOPAEDIA OF
NANO LABS

J. Fox

ANMOL PUBLICATIONS PVT. LTD.
NEW DELHI - 110 002 (INDIA)

ANMOL PUBLICATIONS PVT. LTD.
H.O.: 4374/4B, Ansari Road, Darya Ganj,
New Delhi-110 002 (India)
Ph.: 23278000, 23261597
B.O.: No. 1015, Ist Main Road, BSK IIIrd Stage
IIIrd Phase, IIIrd Block,
Bangalore - 560 085 (India)
Visit us at: www.anmolpublications.com

Encyclopaedia of Nano Labs

First Published, 2006
ISBN 81-261-2977-8

PRINTED IN INDIA

Printed at Mehra Offset Press, Delhi.

Contents

Preface

This publication titled "Encyclopaedia of Nano Lab" provides details and guidelines about the nanotechnology labs around the work and areas related to the intellectual property. Most of the *hottest* nanotechnology labs in various universities which are famous for nano science and technology research. More often, the work undertaken in these labs foucs on commercially develop technology from the lab, as per cleints's requirement as nanoscale materials, deveices and sysems. These Nano Labs has been responsible foremost promising and discruptive breakthroughs the nanotechnology & nanoscale There are so many other applications of nano chemistry ranging from mundane to incredibly profound. Other applications include energy storage materials, photovoltaic cells, fuel cells, polymer nanoclabs, super-tough nanocoatings, landmine detectors, and thousands more. Although it is a field that has already seen its share of hype, there is a general consensus that over the long term, nano labs is likely to reconfigure everything. Nanostructured materials do not represent a new phenomenon. However, the ability to probe, manipulate, understand and engineer matter at atomic scales has only recently come within our grasp. Developments such as the invention of the Scanning Tunneling Microscope in 1981 have since made nanoscale science a reality. Today, the term nanotechnology is used to describe specific construction and manipulation performed on an atomic and molecular scale. Generally, systems with a maximum size of less than a hundred nanometers are classified as nanotechnology. Nanotechnology is a completely new production process in a molecular world, opening up possibilities far beyond what we have known before. Nature provides the inspiration for this new technology – from time immemorial, nanomachines have been at work in all organic cells, building our world of plants, animals and people from a store of chemical elements. Experts world over believe nanolab will have a more far-reaching impact than the silicon integrated chip with many more applications than electronics. Nanotechnology is expected to revolutionize certain areas, such as semiconductors, pharmaceuticals and materials. The total societal impact of nanotechnology is expected to be much greater than that of the silicon integrated circuit because it is applicable in many more fields than electronics. Nanotechnology experts believe that because of nano chemistry, we would see more changes in the next 30 years than we saw in the entire last century. This book tends to cover the aforementioned subject areas in their broadest possible terms. All useful supplementary

research and reference tools have been included for readers' further investigations in the subject area of Nano chemistry. This book mainly focuses on the following areas: Nanotechnology; Nanoparticles, Nanomaterials and Nanostructures. This book seeks to explain this complex field of study. It also seeks to provide an understanding of what is happening now in the field of nano chemistry, as well as what to expect in future by generations to come.

—Editor

1

Nano Lab: An Introductory Overview

Nanoscience and Technology at US - DOE Labs

The Role of DOE in the National Nanotechnology Initiative:

"The DOE has a stunning portfolio of research and scientific user facilities devoted to visualizing, characterizing, and controlling the nanoworld - from atoms and molecules to bulk materials - which makes the Department's research capabilities unique in the world. The DOE is currently making a broad range of contributions in these areas. For example, the enhanced properties of nanocrystals for novel catalysts, tailored light emission and propagation, nanocomposites and supercapacitors are all being explored. Nanocrystals and layered structures offer unique opportunities for tailoring the optical, magnetic, electronic, mechanical and chemical properties of materials, and DOE researchers are have synthesized layered structures for electronics, novel magnets, and surfaces with tailored hardness.

Specific examples of past accomplishments at DOE include:

- Addition of aluminum oxide nanoparticles that converts aluminum metal into a material with wear resistance equal to that of the best bearing steel
- Novel optical properties of semiconducting nanocrystals that are used to label and track molecular processes in living cells
- Nanoscale layered materials that can yield a four-fold increase in the performance of permanent magnets
- Layered quantum well structures to produce highly efficient, low-power light sources and photovoltaic cells
- Novel chemical properties of nanocrystals that show promise as photocatalysts to speed the breakdown of toxic wastes

- Meso-porous inorganic hosts with self-assembled organic monolayers that are used to trap and remove heavy metals from the environment

The DOE also maintains a large array of major national user facilities that are ideally suited to nanoscience discovery and to developing a fundamental understanding of nanoscale processes. Large computational facilities at DOE will also be key contributors in nanoscience discovery, modeling and understanding.

National Nanotechnology Initiative Research at the DOE:

Major new efforts in nanoscale science, engineering, and technology at the Department of Energy will take advantage of opportunities afforded by recent advances. These efforts will be part of the Basic Energy Sciences (BES) program and have the following broad goals:

1. to attain a fundamental scientific understanding of nanoscale phenomena, particularly collective phenomena;
2. to achieve the ability to design and synthesize materials at the atomic level to produce materials with desired properties and functions;
3. to attain a fundamental understanding of the processes by which living organisms create materials and functional complexes to serve as a guide and a benchmark by which to measure our progress in synthetic design and synthesis; and
4. to develop experimental characterization tools and theory/modeling/ simulation tools necessary to drive the nanoscale revolution.

The principal missions of DOE in science, energy, defense, and environment will benefit greatly from developments in these areas. For example, nanoscale synthesis and assembly methods will result in significant improvements in solar energy conversion; more energy-efficient lighting; stronger, lighter materials that will improve efficiency in transportation; greatly improved chemical and biological sensing; use of low-energy chemical pathways to break down toxic substances for environmental remediation and restoration; and better sensors and controls to increase efficiency in manufacturing."

- Massachusetts Institute of Technoogy (MIT) - Microsystems Technology Laboratories - explores microsystems, microfabrication, and manufacturing technologies.
- Princeton University - NanoStructures Laboratory - explores and develops new nanotechnologies to fabricate structures; innovates nanoscale electronic, optoelectronic, and magnetic devices.
- New York University (NYU) - Seeman Lab - investigating unusual DNA

molecules in model systems that use synthetic molecules, with a major effort in DNA nanotechnology.

- University of Cincinnati - Nanoelectronics Laboratory - research at or around the nanometer scale in semiconductors and other electronic materials.
- Academia Sinica - Surface Nanostructure Laboratory - creates, characterizes, and exploits nanostructures (1~50 nm) on solid surfaces.
- Waseda University - Ohdomari Lab - research on the process of nanostructure fabrication, and the physics of silicon surfaces and interfaces.
- University of Pune - Scanning Tunneling Microscopy Group - performs research on nanoparticles, nanolithography, and noise in STM.

NIH Nanotechnology and Nanoscience Information

This page contains information on (1) currently active NIH and BECON research and training opportunities and (2) listings of funded grants for NIH and BECON program announcements related to nanotechnology and nanoscience. Additional information on the NIH bioengineering funding opportunities can be found at http://www.becon.nih.gov/becon_funding.htm.

While much of biology is grounded in nanoscale phenomena, NIH has not re-classified most of its basic research portfolio as nanotechnology. Only those studies that use nanotechnology tools and concepts to study biology; that propose to engineer biological molecules toward functions very different from those they have in nature; or that manipulate biological systems by methods more precise than can be done by using molecular biological, synthetic chemical, or biochemical approaches that have been used for years in the biology research community, are classified as nanotechnology projects. NIH participates in the National Nanotechnology Initiative (www.nano.gov), the U.S. federal multi-agency R&D program for nanoscale science, engineering, and technology.

A list of technical and financial contacts for Nanoscience and Nanotechnology related projects is given at: http://www.becon.nih.gov/nano_contacts.htm.

A list of technical and financial contacts for PA-02-125, Bioengineering Nanotechnology Initiative (SBIR) is given at: http://www.becon.nih.gov/nano_contacts_sbir.htm.

Nanomedicine NIH Roadmap Initiatives:

- Nanomedicine Roadmap Homepage

NIH Reports Related to Nanotechnology:

- NIBIB/DOE Workshop on Biomedical Applications of Nanotechnology, March 17-18, 2005

- Nanoscience and Nanotechnology Symposium - June 25-26, 2000
- NHLBI Nanotechnology Working Group Report
- NCI Cancer Nanotechnology Brochure
- Recent Slideshow on NIH Nanotechnology Funding Opportunities/Examples

Other Trans-NIH Initiatives:

- Nanoscience and Nanotechnology in Biology and Medicine
- Bioengineering Nanotechnology Initiative
- Bioengineering Research Partnerships
- Exploratory/Developmental (R21) Bioengineering Research Grants

National Heart Lung and Blood Institute

- Innovative Technologies for Engineering Small Blood Vessels
- NHLBI Programs of Excellence in Nanotechnology

National Cancer Institute

- NCI Alliance for Nanotechnology in Cancer (including current funding opportunities)
- Innovative Technologies for Molecular Analysis of Cancer
- Application of Emerging Technologies for Cancer Research

National Institute of General Medical Sciences

- Single Molecule Biophysics and Nanoscience - Research on, and development of, new and improved instruments, methods, and technologies for nanoscience, and for the analysis of single protein and nucleic acid molecules and their complexes in vivo and in vitro. Current approaches include optical and fluorescent spectroscopies, scanning probe microscopy, and biomechanical techniques to analyze the behavior and heterogeneity of single molecules and subcellular structures at the nanometer scale. Examples of targets for study include protein or RNA folding, enzyme catalysis, signaling, molecular machines, and the assembly and dynamics of complex cellular structures. A major goal is to develop and enhance existing methods and reagents for the 3-D visualization of cellular processes in living cells in real time at high resolution.

National Toxicology Program (NTP)

Three agencies form the core of the NTP:

- National Institute of Environmental Health Sciences of the National Institutes of Health (NIEHS/NIH)
- National Institute for Occupational Safety and Health of the Centers for Disease Control and Prevention (NIOSH/CDC)
- National Center for Toxicological Research of the Food and Drug Administration (NCTR/FDA)
- The NTP is conducting studies on the toxicology of nanoscale materials.
- Report of the first nanotoxicology workshop http://www.nanotoxicology.ufl.edu/ aimed at developing appropriate experimental strategies for evaluating the safety/toxicity of nanomaterials

Additional NIH Nanoscience and Nanotechnology Research Funding Opportunities:

- DEVELOPMENT OF CELL-SELECTIVE TOOLS FOR STUDIES OF THE BLADDER, PROSTATE, AND GENITOURINARY TRACT - Application Receipt Dates: February 1, 2005
- NOVEL APPROACHES TO CORNEAL TISSUE ENGINEERING - Application Receipt Dates: February 1, June 1, October 1, ending June 1, 2005
- NINDS EXPLORATORY/DEVELOPMENTAL PROJECTS IN TRANSLATIONAL RESEARCH - Application Receipt Dates: February 1, June 1, October 1, ending July 31, 2005
- SMALL BUSINESS BIODEFENSE PROGRAM - Application Receipt Dates: April 1, August 1, December 1, ending Aug. 2, 2005
- CUTTING-EDGE BASIC RESEARCH AWARDS (CEBRA) - Application Receipt Dates: February1, June 1, October 1, ending Oct. 5, 2005
- NEUROTECHNOLOGY RESEARCH, DEVELOPMENT, AND ENHANCEMENT - Application Receipt Dates: February 1, June 1, October 1 ending October 2006

Information on Funded Grants and Projects for NIH/BECON Program Announcements:

- Summary of Funded NIH Bioengineering Nanotechnology Initiative (SBIR)
- Summary of Funded NIH Bioengineering Nanotechnology Initiative (SBIR)
- Summary of Funded NIH Bioengineering Nanotechnology Initiative (SBIR)
- Summary of Funded NIH Bioengineering Nanotechnology Initiative (SBIR)

- Summary of Funded NIH Bioengineering Nanotechnology Initiative (SBIR)
- Summary of Funded Nanoscience and Nanotechnology in Biology and Medicine (PAR-03-045)

NSF - NSES

The potential impact of research on the nanoscale is well recognized, and yet to remain nationally and internationally competitive formidable challenges must be met. Through a unique partnership of academic institutions, industrial partners, and national labs, the NSF Nanoscale Science and Engineering Center (NSEC) for Integrated Nanopatterning and Detection Technologies, headquartered at Northwestern University under the leadership of Chad A. Mirkin, is helping to meet these challenges by providing a cross-disciplinary research program well integrated with innovative educational outreach programs.

The NU-NSEC brings together recognized leaders from Northwestern University, University of Chicago, University of Illinois/Urbana-Champaign, and Argonne National Laboratory, who, building on the nanopatterning tools and synthetic methods unique to this Center, are making advances in the development of advanced nanopatterning techniques and nanoscale sensors. Potential applications in disease and biological/chemical detection systems arena are expected to be profound. However, this research will also directly impact new technological directions outside of the existing purview of the Center including molecular electronics, catalysis, information storage, and therapeutics.

In a field as dynamic as nanotechnology, collaborations are a vital means to transfer knowledge and an important component of the Center mission. NSEC researchers have numerous meaningful collaborations with industry, national laboratories, and academia. Additionally, the Nanotechnology Corporate Partners (NCP) program provides a forum for companies and business groups with allied interests to develop partnerships with the NSEC and help transition its technology into the private sector. Working with the Kellogg School of Management, the NSEC also offers the Small Business Evaluation and Entrepreneurs (SBEE) program, which gives scientists and engineers with new technologies assistance with development of comprehensive business plans for presentation to potential investors. A $10 million seed fund was established by Lurie Investment Fund to initiate business development based on technology developed at the Center, and three companies have already been successfully launched.

Excellent in-house user facilities are available to NSEC researchers. Additionally, the State of Illinois committed $5,000,000 to support the development of next-generation instrumentation and to establish a multi-user, multi-purpose, open-access Nanoscale Imaging, Fabrication, Testing and Instrumentation (NIFTI) facility to provide soft lithographic writing, imaging, and analysis capabilities.

The NU-NSEC is also committed to educating the scientists, technicians, and teachers of tomorrow. This commitment is realized not only in the research of graduate students, undergraduates, and postdoctoral associates, but also through the Center's innovative *educational programs.* Some of these programs build on current educational strengths and infrastructure, while others launch new initiatives both internally and through exciting new partnerships with the Museum of Science and Industry, Chicago, and Harold Washington College (a primarily minority institution).

Undergraduates and minority undergraduates from around the country have opportunities to engage in research at the Center. High school science teachers, many from schools with large minority populations, participate in an innovative program which combines hands-on research with curriculum development. NSEC researchers are collaborating with pre-college educators on the development of a "Nanoscience Module" to introduce nanoscience concepts at the high school level through guided hands-on activities and design challenges. The NSEC is also collaborating with the Museum of Science & Industry, Chicago, to produce a design concept for a 13,000 square foot exhibit in nanotechnology, which is expected to open in 2006 and educate millions of museum visitors each year. An interactive web site for children entitled "Mission Spec-tacular" will be launched in spring 2003, and a series of dramatic educational demonstrations entitled "Frontiers of Nanotechnology" has already been premiered with pre-college and lay audiences. NSEC researchers and their students are also actively involved and highly visible in the broader community, through lectures, radio, and television.

Through synergistic research of the highest quality, strong and active collaborations, and innovative outreach programs, the NU-NSEC is advancing scientific knowledge and understanding in the area of integrated nanopatterning and nanoscale detection technologies, while advancing formal education as well as public understanding of nanoscale science and engineering.

Sandia National Labs

It's no wonder that venturing into the tiny domain of atoms and molecules is catching on. Amazing discoveries are being made, inspiring scientists the world over.

Throughout the scientific community, including Sandia National Laboratories, researchers say building things atom-by-atom or molecule-by-molecule will revolutionize the production of virtually every human-made object. *Exciting* prospects—but they also point out the promise of nanotechnology can *only* be realized if we learn to understand the special rules that control behavior at this small scale and develop the skill needed to integrate these concepts into practical devices.

The excitement stems from the understanding that the behavior of materials at the nanoscale is *nothing* like that at the large scale. Only recently have the necessary tools, such as powerful new microscopes, been developed to let researchers see these surprising behaviors. Sandia National Labs' scientists and engineers are among the leading architects and builders of these tools. To appreciate their unique design features, it helps to get an idea of the size of things at the atomic level. In nanoscience objects are measured in nanometers, 1 billionth of a meter. For comparison, the smallest features on current computer chips, measure about 200 nanometers. And a human hair is 100,000 nanometers thick.

Moving beyond observation, scientists are now poised to make exciting advances in nanotechnology, the creation of materials, devices and systems through the *control* of matter at the atomic level. By understanding and controlling the way molecules organize into nanoscale patterns, scientists are discovering new phenomena and learning to design materials with vastly different sets of properties. As one Sandia scientist put it "Design possibilities are limited only by one's imagination."

Sandia Labs continues to enhance its proficiency in many fields, all in keeping with its Department of Energy mission to unite science and engineering to serve national needs. And along with world class capabilities in materials science, micro fabrication - including 40,000 square feet of clean room space - high performance computing and systems engineering - Sandia is uniquely positioned to be the integrating center for new discoveries in nanoscience.

Sandia is uniquely positioned to be the intergrating center for new discoveries in nanoscience. Nano materials are already helping to attain scientist's vision for new technology such as, the *micro-chemlab*, by increasing a thousandfold the ability to collect the chemicals of interest. Currently in the development stage, this handheld device will carry out all of the functions of a full chemistry laboratory allowing first responders, military, and emergency personnel to rapidly access chemical and biological hazards.

All of these advances are coming from the work of multi-disciplanary teams. Their ability to bring together and integrate micro-scale devices is another key part of the nanotechnology success story Through strong commitment and continued partnerships with other national labs, universities and industry. Sandia scientists believe the promise of nanotechnology applications that touch peoples lives can be realized. Grand prospects based on tiny environments. It will require thinking big in a nano sized world.!

U.S. Naval Research Lab: Nanotechnology Science and Technology

Strategic Research Area

- To achieve dramatic, innovative enhancements in the properties and performance of structures, materials, and devices that have controllable features on the nanometer scale (i.e., tens of angstroms).

- The ability to affordably fabricate structures at the nanometer scale will enable new approaches and processes for manufacturing novel, more reliable, lower cost, higher performance and more flexible electronic, magnetic, optical, and mechanical devices.

Recognized Applications of Nanoscience

- Ultrasmall, highly parallel, computers with multi-teraflop speed
- Image information processors, e.g., extraction and recognition
- Low-power personal and autonomous communication and computation devices
- high-density information storage devices, e.g., terabit/cm2 nonvolatile memory
- Lasers and detectors for weapons and countermeasures
- Optical (infrared, visible, ultraviolet) sensors for improved surveillance an targeting
- Integrated sensor suites for chemical and biological agent detection
- Catalysts for enhancing and controlling energetic reactions
- Synthesis of new compounds (e.g., narrow-bandgap materials)
- Designer materials with combinations of properties that do not currently exist

SRA Representatives

Air Force: Dr. Gernot S. Pomrenke
Army: Dr. Henry Everitt
DARPA: Dr. Christie Marrian
Navy: Dr. James Murday (Chair)

US Office of Naval Research

The Surface Nanoscience and Sensor Technology section of the Naval Research Laboratory specializes in:

(1) basic and applied research in surface science, the mechanical properties of materials, the molecular basis of adhesion, film nucleation and growth, and chemical/biological sensors;

(2) development of tip-based probes for studies in the area of nanostructures and nanomechanics of surfaces and single molecules; and (3) development of new surface/interface analysis techniques for structural and chemical analysis of materials.

Current Areas of Research

Structure of semiconductor surfaces and interfaces

We are studying the atomic-scale structure and reactivity of single crystal silicon, germanium, and III-V compound semiconductor surfaces and interfaces in ultra-high vacuum (UHV) using scanning tunneling microscopy (STM) and other surface analysis techniques. Due to the rapidly shrinking size of electronic devices, such studies are vital to the development of future Navy electronics.

Nanomechanics of surfaces

The nanomechanical properties of materials has become an area of increasing importance as the scale of coatings, multilayer films, particles, components and devices approach atomic dimensions. A goal of our nanomechanics program is to replace the phenomenological understanding of adhesion, tribology and interfacial/fracture mechanics with specific atomic- or molecular-scale mechanisms. Our research objectives include measuring the intermolecular and surface forces of materials with molecular specificity and resolution; elucidating the mechanics of materials or species under compression, tension, or shear; and developing new science-based molecular models for adhesion and tribology. This program also focuses on the relationship between interfacial chemistry and the structural and mechanical properties of interfaces by applying state-of-the-art nanoprobe technology including atomic force microscopy (AFM) and nanoindenation techniques.

Interaction forces of biomolecules

Atomic force microscopy (AFM) offers the unique ability to perform nanomechanical measurements on individual biological molecules. We have developed this ability and are using it to investigate inter- and intra-molecular interaction forces in model systems, between complementary strands of DNA, and between antibodies and their antigens. Molecular modeling studies are carried out to provide insights into the bond rupture mechanisms that underly these experiments.

FABS / BARC / FDA biosensors

Building on our research into the interaction forces between biomolecules, we are developing sensors for biomolecules and other chemicals, viruses, and bacteria. These sensors use microfabricated transducers to detect the minute forces with which antibodies bind to particular molecules. This research will result in the development of portable, highly-sensitive, automated chemical and biological sensors. The Force Differentiation Assay (FDA) and the Force Amplified Biological Sensor (FABS) are expected to deliver orders-of-magnitude sensitivity improvements over methods that are now in common use. The Bead Array Counter (BARC) will operate on similar principles but will trade sensitivity for the ability to simultaneously detect hundreds of types of molecules.

Quantitative surface analysis

The surface analytical techniques of x-ray photoelectron spectroscopy (XPS) and Auger electron spectroscopy (AES) are used to quantitatively determine the surface composition or in-depth distribution of low-level dopants in a variety of advanced materials including semiconductors, polymers, diamond films, and engineering alloys. These approaches are employed also for tribology, surface cleaning, and insulation degradation studies. Experimental and theoretical methods have been developed to overcome problems caused by matrix effects, surface segregation, background, peak overlap, specimen charging, sputtering artifacts, etc. Auger lineshapes and factor analysis methods are used extensively to extract chemical information.

Langmuir-Blodgett and self-assembled films

Research is conducted on physicochemical properties and effects of surface-active organic compounds that 1) form organized monomolecular films and 2) can be transferred to solid surfaces as mono- or multilayer films with precisely controlled properties such as packing density and phase state. Such films are also used in fundamental research studies as reference standards for scanning probe microscopes; as model systems to help understand the physical and chemical properties of biological membranes; and as model systems to increase understanding of the molecular basis of adhesion and lubrication. They have also been used in practical applications that take advantage of the unique properties of organized thin films, for example, as rapidly-equilibrating active coatings for gas sensors. Properties and effects of films that occur naturally on the sea surface have also been studied.

Facilities

Scanning probe microscopes

Scanning tunneling microscopy/spectroscopy (STM/S) enables the surface topography, chemical reactivity, and electronic structure of conductive substrates to be observed with atomic-scale resolution. Atomic force microscopy (AFM) can provide nanometer-scale resolution of surface topography, mechanical properties, and tip-surface interaction forces on both conductive and insulating substrates. The tip-surface interaction forces, including frictional forces, can be measured with nanonewton (single chemical bond) resolution.

- Homebuilt ultra-high-vacuum (UHV) STM/S with facilities for low-energy electron diffraction (LEED) and Auger electron spectroscopy (AES), for characterizing atomic-scale surface properties;
- Homebuilt STM/S with fast, variable-temperature STM and facilities for LEED, AES, temperature programmed desorption, for real-time atomic-scale observations of surface reactions;

- Park Scientific Instruments AutoProbe UHV STM/AFM integrated with the NRL Molecular Beam Epitaxy (MBE) Epicenter for characterizing semiconductor surfaces following MBE and related processing;
- RHK Technology STM 2000 UHV STM/S;
- Homebuilt ambient STM/S with capability of generating high-frequency surface acoustic waves (SAW) for measurement of elastic properties of surfaces;
- Nanoscope IIIa multimode AFM (lateral force, magnetic force, and tapping modes) equipped with breakout box and Hysitron TriboScope nanoindentation instrument for mechanical property measurements;
- Nanoscope IIIa multimode AFM (lateral force, magnetic force, and tapping modes) equipped with breakout box and force-volume mapping system;
- Homebuilt ambient AFM for adhesion and tribology studies;
- Homebuilt liquid-cell AFM for molecular recognition studies;
- Topometrix TMX 2000 AFM, belonging to code 6174, but housed within code 6177.

Surface analytical techniques

These techniques are used to characterize the surfaces of solids with respect to their electronic and geometric structure, mass, molecular weight, or optical properties. The techniques include X-ray photoelectron spectroscopy (XPS) and Auger electron spectroscopy (AES). Both approaches carry information about the electronic structure of the solid and its chemical composition. In addition, a rastered electron beam (AES) or parallel imaging (XPS) can produce images of chemical species.

- Perkin-Elmer 660 50 nm scanning Auger microprobe (SAM) with sputter depth profiling capability and X-ray dispersive analyzer, for analyzing thin films and determining the lateral or in-depth distribution by depth profiling of elements at the <0.5 atomic percent level.
- Surface Science Instruments ESCA301 150 µm monochromatic XPS with depth profiling capability and angle-resolved analysis.
- VG Scientific model 220i-XL XPS with monochromatic and non-monochromatic X-ray sources, depth profiling capability, small area analysis (20 µm) and <5 µm spatial resolution parallel image capability.
- JA Woollan M44 410-750 nm ex situ spectroscopic ellipsometer with ambient and liquid cells for film thickness, surface roughness, optical constant, and composition measurements.

Chemistry laboratories

These facilities are primarily intended for the preparation and characterization of 1) antibody or DNA-derivatized surfaces and 2) Langmuir-Blodgett and self-assembled films.

- Biochemistry and molecular biology facilities with -70° C freezer, UV/vis spectrophotometer, PCR cycler, gel box, microplate reader, incubator, glove box, centrifuge, etc.
- Zeiss Axiovert and Axiotech reflected- and transmitted-light optical microscopes with epifluorescence and differential interference contrast (DIC) attachments; equipped with Narashige three-dimensional hydraulic micromanipulator and video microscopy/digital video capture system.
- Dedicated surface chemistry clean-room laboratory with KSV 5000 Langmuir-Blodgett System.

Magnetics equipment

The following equipment for generating and measuring magnetic fields is part of the FABS/BARC/FDA program.

- Group 3 DTM-141 Digital Teslameter with serial communications and MPT-141 miniature Hall probe.
- Walker Scientific 40 kOe pulsed magnetizing coil with 1 inch bore.

Scanning electron microscopy, Fourier transform IR and Raman spectroscopy, solid state NMR, X-ray fluorescence, Rutherford backscattering, MS/MS, and other supporting techniques are also available within the Chemistry Division.

W.M. Keck Nanostructures Laboratory - University of Massachesetts, Ahmerst

The W.M. Keck Nanostructures Laboratory is a part of the Materials Research Science and Engineering Center (MRSEC) and the Silvio O. Conte National Center for Polymer Research at the University of Massachusetts at Amherst. It is equipped with several Atomic Force Microscopes, Optical Microscopes, Variable Angle Spectroscopic Ellipsometer (VASE), Thin Film Optical Profilometer, Stylus Profilometer, Metal Evaporator etc. for the use of UMass community and as well as from other academic institutions and Industry. The mission of the Keck Nanostructures Laboratory is to provide access to material characterization equipment, technical support, training and consultation, as well as to perform a range of services for users in the area of Atomic Force Microscopy (AFM), Small Angle X-ray Scattering (SAXS), Variable Angle Spectrocopic Elliposmetry (VASE) and Optical Microscopy. The laboratory capabilities include high-resolution imaging of materials structure, analysis of X-ray Diffraction patterns, Thin Film characterization etc. In addition the Nanostructures Laboratory has access to the Confocal Laser Scanning Microscope from the Optical Microscope facility of the MRSEC

NanoTech User Facility at the UW (NTUF)

Training

Welcome to the NanoTech User Facility (NTUF). NTUF is available to graduate students and industrial researchers as a cost center. NTUF staffs will provide training on instruments upon request. Please register online and we will contact you to schedule training sessions.

Training consists of two sessions. First, we will provide an in-depth introduction on the principles of the instrument and demonstrate capabilities by running standard samples. The second training session will be hands-on and users are encouraged to work with their own samples. We may request additional practice sessions depending on qualification on the use of equipment.

For qualified users who can perform independent experiments and who have agreed with NTUF policy, NTUF is available 24 hours a day and seven days a week.

- Zeiss Confocal Microscope Guided Tour
- Principles of Confocal LSM
- Beam Path Configuration Guide
- Zeiss LSM 510 Operating Procedure
- Volocity Users Guide
- Digital Multimode III Operating Procedure
- Nanoscope III User Manual (© Digital Instruments)
- Nanoscope III schematic (© Digital Instruments)
- Leica Operating Procedure
- Quick Procedure for Raman Spectroscopy
- Raman system schematic
- FEI Sirion Operating Procedure
- E-beam Lithography Training Procedure I
- E-beam Lithography Training Procedure II
- E-beam Lithography Introduction
- E-beam Resist (PMMA) Data Sheet (© MicroChem)
- NTUF E-Beam Resist Information

- NTUF Master Template I
- NTUF Master Template II

Charles B. Musgrave

Quantum Simulations of Molecular Processes in Chemical Engineering

Our research program focuses on using quantum mechanics to simulate molecular processes of important engineering problems. Our approach is fundamental and interdisciplinary and combines quantum mechanics, chemical kinetics, surface chemistry and materials chemistry. Our goals are to develop a fundamental molecular and mechanistic understanding of the processes underlying important new technologies and to establish quantum chemical simulations as a practical engineering tool for computational prototyping of systems with molecular detail. Systems we investigate include atomic layer deposition, dye sensitized nanostructured solar cells, solid oxide fuel cells, hybrid interfaces between semiconductors and organic and biological molecules, molecule-surface electrical junctions and nanotechnology.

Atomic Layer Deposition

ALD is a process that has recently gained enormous attention. It involves exposure of surfaces to alternating pulses of different precursors to form atomic layers of deposited material through self-limiting surface reactions. Currently, the most active area of ALD research is the deposition of high-K gate dielectrics, such as ZrO_2 and HfO_2 because finding a replacement for SiO_2 gate dielectrics is the most critical technical challenge facing the semiconductor industry in the current transition to nanoelectronic devices. Although research on ALD of high-K materials has recently exploded, no detailed mechanisms or first principles studies had been reported until we reported detailed mechanisms for ALD of several important materials including ZrO_2, HfO_2,, HfN, Al_2O_3, SiO_2 and Si_3N_4. In addition to being the first group to describe detailed mechanisms of ALD, we are creating the fundamental framework for the understanding of the ALD process which provides a based upon which ALD can be extended and improved. We are also exploring reactions for low-temperature ALD on organics and self-assembled monolayers for area-selective ALD and molecular electronics, ALD for fuel cells, and dye sensitized solar cells.

Dye-Sensitized Solar Cells

The heart of dye-sensitized solar cells is the dye-molecule bound to a metal oxide nanostructured surface who's primary function is to covert photons to electron hole pairs in order to harvest solar energy. The goal of our work on dye-sensitized solar cells (DSC) is to develop new molecular technologies to control the key interfacial properties of DSC with the aim of improving their efficiency, and reducing their cost and the energy required for their manufacture. Our strategy is to use quantum chemical simulations and carrier transport simulations to perform molecular engineering of the DSC interface structures by computationally prototyping both the interfacial physics which governs the operation of the DSC as well as the chemical processes involved in their fabrication. The DSC has great potential for economically harvesting significant amounts of the solar energy incident upon the earth's surface without the production of green house gases. Furthermore, despite being early in their development cycle relative to traditional photovoltaics, much progress has been made in both the fabrication and efficiency of the DSC with efficiencies exceeding 10%.

Our view of DSC is that the surface-tethered dye is an optoelectronic molecular device whose function critically depends on the quantum electronic structure. From our previous work in organic functionalization of semiconductors surfaces and ALD of metal oxides we have become one of the world's leading groups in the electronic structure of organic functionalized surfaces as well as in the electronic structure of organometallic molecules on metal oxide surfaces and we leverage this expertise to enable us to expand the theory and understanding of the DSC with the hope that this will lead to improvements and subsequent faster implementation of DSC technology.

Low Temperature Solid Oxide Fuel Cells

Fuel cells have the potential to play an immensely important role in a potential hydrogen economy. Solid oxide fuel cells (SOFC) use solid electrolyte gates and have many advantages over polymer gate systems. The main disadvantage is that they must be operated at very high temperatures (1000-1200 C). Our goal is to simulate oxygen ion migration through the electrolyte to determine the migration mechanism and to design and computationally prototype new electrolytes with higher ion conductivities with the goal of reducing the operating temperature of the fuel cell by using dislocations, and built-in chemical potential gradients, and ultra thin membranes fabricated using ALD.

Hybrid Bio and Organic Semiconductor Devices

Semiconductors are at the center of many key technologies including computation, communication, data storage and sensing. Inorganic interfaces in semiconductor based devices have been key components in the unique properties that these devices exploit for their operation. We have undertaken an extensive study of how various organic molecules react on semiconductor surfaces and developed the most comprehensive theoretical description of organic reactions on semiconductors. We are extending this leadership position in two ways. First, by moving from studying the reactivity of organics on semiconductors to investigating the unique electronic properties of the interfaces formed between various organic functional groups and semiconductor surfaces. Second, by extending the types of organic molecules on semiconductors we study to biological molecules, such as amino acids. We have recently found several interesting new types of reactions between biological molecules and semiconductor surfaces which are not only unique in their reactivity, but which also exhibit strong quantum mechanical coupling between the biological molecule and the semiconductor, potentially leading to new types of electrical junctions in molecular electronics. We are also exploring the use of this phenomenon to electrically connect carbon nanotubes to each other.

Molecular Electronics

There has been significant recent interest in using molecules as components in electronic circuits. Critical to this area is the ability to transport electrons between molecules and bulk materials across the interface formed between the molecule and the material surface. This is mostly governed by the electronic structure of the molecule-material interface. Several of our projects involve charge transfer between molecules and surfaces including dye-sensitized solar cells, the oxidation and reduction in fuel cells and our quantum resonance junctions between certain amino acids and Si and Ge surfaces. Our strategy is to use quantum simulations to develop a clear understanding of the charge transfer process so that we may engineer molecules to optimize them as electronic junctions.

Theoretical Studies of a Hydrogen Abstraction Tool for Nanotechnology

Processes that use mechanical positioning of reactive species to control chemical reactions by either providing activation energy or selecting between alternative pathways will allow us to construct a wide range of complex molecular structures. An example of such a process is the abstraction of hydrogen from diamond surfaces by a radical species attached to a mechanical positioning device for synthesis of atomically precise diamond-like structures. In the design of a nanoscale, site-specific hydrogen abstraction tool, we suggest the use of an alkynyl radical tip. Using *ab initio* quantum-chemistry techniques including electron correlation we model the abstraction of hydrogen from dihydrogen, methane, acetylene, benzene and isobutane by the acetylene radical. Of these systems, isobutane serves as a good model of the diamond (111) surface. By conservative estimates, the abstraction barrier is small (less than 7.7 kcal mol^{-1}) in all cases except for acetylene and zero in the case of isobutane. Thermal vibrations at room temperature should be sufficient to supply the small activation energy. Several methods of creating the radical in a controlled vacuum setting should be feasible. Thermal, mechanical, optical and chemical energy sources could all be used either to activate a precursor, which could be used once and thrown away, or alternatively to remove the hydrogen from the tip, thus refreshing the abstraction tool for a second use. We show how nanofabrication processes can be accurately and inexpensively designed in a computational framework.

COLLEGES, UNIVERSITIES, LIBRARIES & INSTITUTIONS

- NANO.DOT, http://nanodot.org/ —A Foresight Institute website—News & Discussion
- Nanotechnology White Paper Library, http://nanobusiness.org/whitepaper.html—The NanoBusiness Association's "Nanotechnology White Paper Library" is a collection of links to more than 500 nanotechnology white papers, presentations and reports. Most are free; some may require registration with their host website.
- CRN, http://www.crnano.org/—Center for Responsible Nanotechnology
- The Institute of Nanotechnology, http://www.nano.org.uk/—The Institute of Nanotechnology has been created to foster, develop and promote all aspects of science and technology in those domains where dimensions and tolerances in the range of 0.1 nm to 100 nm play a critical role
- UCSC Engineering, http://www.cse.ucsc.edu/research/vision.html—University of California, Santa Cruz / Baskin School of Engineering
- Nanotechnology Initiative Research Group, http://www.me.berkeley.edu/nti/home.html—Department of Mechanical Engineering University of Berkeley
- Stanford Nanotechnology, http://scpd.stanford.edu/SCPD/courses/contentView/nanotechnology/—Stanford University is at the forefront in multidisciplinary research and innovation in nanoscience and nanotechnology. The Stanford Center for Professional Development, in its commitment to extend Stanford research and teaching to students in industry, has developed this portfolio of courses and programs in this exciting new field.
- NSL NanoStructures Laboratory, http://nanoweb.mit.edu/ —The Nanostructures Laboratory (NSL) at MIT develops techniques for fabricating surface structures with feature sizes in the range from nanometers to micrometers, and uses these structures in a variety of research projects. The NSL is closely coupled to the Space Nanotechnology Laboratory (SNL) with which it shares facilities and a variety of joint programs.
- MIT Stanford UC Berkeley Nanotechnology Forum, http://www.quantuminsight.com/MSB/organization.html—The MIT Stanford UC Berkeley Nanotechnology Forum is an all-volunteer organization. It is organized under the MIT Club of Northern California and has affiliations with MIT, Stanford University, and the University of California at Berkeley. The Forum is run by the Chairman and is guided by the Steering Committee.

- The Foresight Institute—Non-profit institute focused on nanotechnology, the coming ability to build materials and products with atomic precision, and systems to aid knowledge exchange and critical discussion, thus improving public and private policy decisions.
- Nanotechnology Now—Comprehensive and informative nanotechnology portal. Provides introduction to nanotechnology, general information, images, interviews, news, events, research, books, glossary and links.
- Applied Computational Research Society—Nonprofit organization fostering interdisciplinary approaches to modeling and simulation in the advancement of computational science and high technology. Focused on nanoscience and MEMS applications.
- The ASME Nanotechnology Institute—American Society of Mechanical Engineers: nanotechnology resources, events, programs, activities, online courses, webcasts, news and links.
- CMP Cientifica—News, networks, conferences and nanotechnology resources in Europe
- Euspen—European Society for Precision Engineering and Nanotechnology—contains society news, conference information, discussion forums and expertise search. Euspen is a not-for-profit company based at Cranfield University in the UK, with members around the world.
- Exploring the Nanoworld—Movies of nano-structured materials: ferrofluids, memory metals, amorphous metals, LEDs, re-entrant foams, self assembly, DNA, Magnetic Resonance Imaging, atomix imaging, Lego models, Giant Magneto Resistance, silicon, carbon nanotubes, polymers, and semiconductor devices. National Science Foundation supported Materials Research Science and Engineering Center on Nanostructured Materials and Interfaces at the University of Wisconsin-Madison.
- The gRobots Project—An open source project to build and run a globally distributed supercomputer for the simulation of evolvable nanotechnology.
- How Stuff Works: How Nanotechnology Will Work—Animated narrative shows how Nanotechnology has the potential to totally change manufacturing, health care and many other areas.
- IBM Almaden STM Molecular Art—Some of the famous images of atoms and molecules made with IBM's scanning tunneling microscope.
- Institute for Atomic-Scale Engineering—Personal page of Forrest Bishop: published and unpublished concept articles.

- Institute of Nanotechnology (UK)—The Institute of Nanotechnology has been created to foster, develop and promote all aspects of science and technology in those domains where dimensions and tolerances in the range of 0.1 nm to 100nm play a critical role.
- Iran Nanotechnology Policy Studies Committee—A committee for studying nanotechnology trends in Iran and the world.
- Minatec—Centre for innovation in micro and nanotechnology. Teaching, research, and industrial development.
- Minatec—Centre for Innovation in Micro and Nanotechnology in Grenoble. Teaching, research, and industrial development.
- MITRE Corp. Nanosystems Group—Reviews current nanotechnology research and people behind it, downloadable copies of several widely-used articles. Nanoelectronics focus.
- Molecular Manufacturing Shortcut Group—Promotes developing nanotechnology as a way to facilitate space exploration and settlement.
- Moletronics and Nanotechnology references—Links to sites with information mainly on molecular electronics.
- Nano Science & Technology Institute—Events, publications and news in nanotechnology, microtechnology and biotechnology.
- Nano Space—Nano technology as it applies now and in the future to Space Exploration and Studies.
- NanoApex—Nanotechnology portal.
- Nanobase—Loyola College, Maryland. National Science Foundation sponsored site presenting up-to-date sources of information on nanotechnology in the areas of major research centers, funding agencies, major reports and books.
- NanoComputer Dream Team—Multidisciplinary project to make Earth's first nanometer supercomputer via Internet.
- Nanodot—News and Discussion of Coming Technologies: Weblog devoted to nanotechnology-related news stories, with attached discussion forum. (based on Slashdot software)
- Nanoforum.org: European Nanotechnology Gateway—Offers background information, news and developments in industry and research and a selection of publications.
- nanogloss—Online dictionary of nanotechnology.
- Nanoguys—Nanotechnology Career Consultants. Recruiting and placement

of staff in nanotechnolgy related positions such as Organic, Inorganic, Biological, Project Management, Research Scientist, Engineer, and Lab Technician.

- Nanomagazine—Interviews with people in the Nanotechnology field.
- NanoManipulator: 3rdTech—Commercial version of the NanoManipulator system originally from University of North Carolina, Chapel Hill.
- NanoManipulator: University of North Carolina—The nanoManipulator provides an improved, natural interface to SPMs (STMs and ATMs) by coupling the microscope to a virtual-reality interface giving a virtual telepresence on the surface, scaled by a factor of about a million to one.
- Nanomedicine—Devoted to the medical applications and implications of molecular nanotechnology. Nanomedicine may be defined as the monitoring, repair, construction and control of human biological systems at the molecular level, using engineered nanodevices and nanostructures. Includes the Nanomedicine Art Gallery. (From the Foresight Institute.)
- Nanoscale Science and Engineering—National Science Foundation. Activities, Information & Education, Research & Development, Connections, and NSF & NNI (National Nanotechnology Initiative) Reports.
- NanoScout—Portal with forums, and links sorted by topic and popularity.
- NanoSpace—The Center for NanoSpace Technologies is a Texas-based nonprofit scientific research and education foundation chartered to conceive, establish, and conduct cutting-edge technology, research, and development in the areas of aerospace, education, energy, life sciences, and shipping & transportation.
- NanotechNews—News for nanotechnologists and investors: regular updates, many links to other nanotechnology websites, archives, search function, newswire.
- Nanotechnology—Institute of Physics quarterly journal
- Nanotechnology—Institute of Physics monthly journal for aspects of nanoscale science and technology
- The NanoTechnology Group—A consortium of nano companies, universities and organizations developing a nano science curriculum for K-12.
- Nanotechnology in Manufacturing—What Next? The Coming Revolution In Manufacturing: Autodesk Technology Forum Presentation by John Walker, May 10th, 1990. Somewhat dated but well worth reading.
- Nanotechnology Journals Database—References, and links to, papers about

nanotechnology, from selected scientific journals (eg PRB, PRL, JACS, JAP, JPCB).

- Nanotechweb.org—A portal for nanoscience and nanotechnology online resources for academics, industrialists and investors, as well as for laymen interested in nanotechnnology.
- NanoTechWire.com—Provides information for nanotechnology and research.
- Nanotek—Nanotechnology: articles online, news, pictures, cvs
- nanoTEN—Nanotechnology consulting firm. Services include nanotechnology matchmaking, assessment of trends, feasibility studies, materials and products analysis, and presentation assistance.
- nanoTITAN—Nanotechnology Software, Data & Services: nanoTITAN offers Java libraries and applications, a nanodevice database, the nanoML specification, custom simulation and visualization to the nanotechnology and general scientific communities.
- Nanotube Patent Search—Results from a search of nanotube or nanotube related patents at the United States Patent & Trademark Office
- The Nanotube Site—Michigan State University's Library of Links for the Nanotube Research Community
- NanoWave—An MIT startup, Nanowave offers SPPE (Scanning Probe Position Encoder) technology for picometer-order (subnanometer) resolution over ranges greater than several inches.
- National Science Foundation—Partnership in Nanotechnology (US). Recent technical and policy reports, funding opportunities, profiles of projects currently supported, and links to other nanoscience/technology sites.
- Phantoms—European Union sponsored nano electronics network. Contains comprehensive information about European nanotechnology, and links to groups and events worldwide.
- Polymer Simulation—Computer simulation of a polymer sticking to a surface.
- Scanning Tunneling Microscope (STM)—Describes the invention of the topografiner, a precursor instrument, between 1965 and 1971, and also tells of the STM's development.
- Scientific American—Nanotechnology articles, some free and some archived.
- Self Replicating Systems—NASA and Self-Replicating Systems: Implications for Nanotechnology.
- Small Times—Daily articles covering MEMS, nanotechnology, and microsystems, with a business angle.

- Stony Brook Buckyballs—A virtual tour of fullerenes in Laszlo Mihaly's laboratory at the Physics Department at SUNY, Stony Brook.
- STT: Netherlands Study Center for Technology Trends—Somewhere in this confusing site is good material in English, though it is often labeled in Dutch, and Dutch material is often labeled in English. STT hopes to bring their project to the attention of a wide audience, thereby contributing to the development of nanotechnology.
- Technology Review: Nanotechnology and Materials—Articles, discussion and services for nanotech built on MIT's award-winning monthly magazine.
- Thermodynamics Research Laboratory—Classical, Quantum, Nanoscale & Statistical Thermodynamics & Mechanics data and property calculation sites.
- Virtual Journal of Nanoscale Science & Technology—A weekly multijournal compilation of the latest research on nanoscale systems. Published by the American Institute of Physics and the American Physical Society.
- VISION Online—Virtual Institute supported by the EC, committed to promotion and training in the precision engineering and nanotechnologies. Contains discussion forums, training programmes and patent library.
- Xerox/Zyvex Nanotechnology—Former Xerox Palo Alto Research Center Nanotechnology page: brief introduction to core concepts of molecular nanotechnology (MNT), and links for further reading.
- Yahoo Groups: Nanotechnology—Active nanotechnology discussion group.
- Yashnanotech—Provides marketing and consultancy services for products developed by nanotechnology. Promotes nanotech awareness through training, educational publications, seminars, workshops and conferences.
- Albany NanoTech "... a fully-integrated research, development, prototyping, pilot manufacturing and education resource managing a strategic portfolio of state-of-the-art laboratories, supercomputer and shared-user facilities and an array of research centers located at the University at Albany—SUNY."
- Arizona State University (ASU) Graduate Research Training Program in Biomolecular Nanotechnology
- Arizona State University (ASU) Center for Applied NanoBioscience
- Arizona State University (ASU) Biodesign Institute
- Arizona State University (ASU) Lindsay Lab—Research in Biophysics, Molecular Electronics and Condensed Matter Physics

- Arizona State University (ASU) Center for BioOptical Nanotechnology
- Arizona State University (ASU) Nanostructures Research Group—Center for Solid State Electronics Research
- Asian Institue of Technology Nanotechnology Studies. Bangkok, Thailand
- Australian National University Research School of Physical Sciences and Engineering
- Australian National University Nanotechnology and Mesoscale Physics
- Ball State University Center for Computational Nanoscience
- Baylor College of Medicine Molecular Biology programs
- Beckman Institute Biological Sensors Group (BSG)
- Berkeley Nanotechnology Club (BNC)
- Brigham Young University Cleanroom A class 10 cleanroom located on the BYU campus, used for research in solid state, optoelectronics, photonics, nanotechnology, microfabrication, MEMS, micro-fluidics, etc.
- Boston University College of Engineering Departmental program in Nanotechnology. "... is concerned with the mechanical and electronic properties of nanometer scale systems as well as macroscopic systems that require nanometer tolerances in fabrication, control, and measurement."
- Boston University Laboratory for Nanometer Scale Engineering
- Boston University Mohanty Group. "The central theme of our research is the exploration of quantum mechanical effects in engineered nanoscale structures and devices with a goal to study fundamental physical phenomena."
- Brown University Nano & Micromechanics Laboratory, part of the Mechanics of Solids and Structures Group in the Division of Engineering
- California Institutes for Science and Innovation: California NanoSystems Institute (CNSI) University of California at Los Angeles and University of California at Santa Barbara California Institute for Telecommunications and Information Technology (Cal-(IT)2) University of California at San Diego and UC Irvine Center for Information Technology Research in the Interest of Society (CITRIS) University of California at Berkeley, University of California at Davis, University of California at Merced, University of California at Santa Cruz Institute for Bioengineering, Biotechnology and Quantitative Biomedical Research (QB3) University of California at Berkely, UC San Francisco, UC Santa Cruz
- California Institute of Technology (Caltech) Materials and Process Simulation Center

- Caltech Roukes Group—Explorations with Three-Dimensional Nanostructures in Physics, Engineering and Biology
- Caltech Kavli Nanoscience Institute (KNI)
- Caltech Solid State Device Physics (SSDP) Research Group. Department of Applied Physics.
- Carnegie Mellon University Nanomaterials for Magnetic, Electronic and Optical Applications
- Carnegie Mellon University Matyjaszewski Group—Research—Organic/Inorganic Nanocomposites
- Center for Applied Research and Technology at CMU The Central Michigan University Research Corporation (CMURC) is a not-for-profit organization established to facilitate innovative research and development opportunities between the university and high technology companies.
- Chalmers University of Technology, Sweden Nanoscale Science
- Chalmers University of Technology, Sweden Microtechnology Centre
- Chiang Mai University Nanomaterials Research Unit
- City College of New York New York Center for Biomedical Engineering
- City College of New York Molecular Beam Epitaxy (MBE) Program
- City University of New York IGERT: Nanostructural Materials & Devices (Integrative Graduate Education and Research Traineeship Program)
- City University of Hong Kong Center Of Super-Diamond and Advanced Films (COSDAF)
- Clarkson University Center for Advanced Materials Processing.
- Clemson University COMSET—Center for Optical Materials Science and Engineering Technologies.
- Clemson University Center for Advanced Engineering Fibers and Films (CAEFF).
- Clemson University Rao Group.
- Columbia University Center for Electronic Transport in Molecular Nanostructures NSF Funded "Nanoscale Science and Engineering Centers" Real Video (2:50) Professor James Yardley speaking about Nanotechnology
- Columbia University Materials Research Science and Engineering Center (MRSEC)
- Columbia University Environmental Molecular Science Institute
- Columbia University NANOWEB—Nanotechnology Resources at Columbia University

- Columbia University Adams Group
- Columbia University Brus Group
- Columbia University Herman Group—Laser Diagnostics & Solid State Physics
- Columbia University Nuckolls Laboratory
- Cornell University Nanofabrication Facility. Part of the National Nanotechnology Infrastructure Network (NNIN)
- Cornell University Craighead Research Group. "... research focuses on creating nanoscale devices..."
- Cornell University Alliance for Nanomedical Technologies
- Cornell University Center for Nanoscale Systems NSF Funded "Nanoscale Science and Engineering Centers"
- Cornell University Nanobiotechnology Center
- Cornell University Molecular BioEngineering Lab
- Cornell University Ober Group Research
- Cornell University Center for Materials Research
- Cornell University Laboratory of Atomic and Solid State Physics (LASSP)
- Cornell University McEuen Group—LASSP
- Cornell University Applied Physics—Nanoscience and Nanotechnology
- Cornell University Applied Physics—Condensed Matter, Solid State, and Materials Research
- Cornell University Applied Physics—Optical Physics, Quantum Electronics, and Photonics
- Cranfield University Nanotechnology Group, Advanced Materials Department
- Curtin University of Technology Nanochemistry Research Institute
- Dalhousie University Zwanziger Lab
- Darmstadt University of Technology Institute of Materials Science—Thin Films Division
- Dartmouth Molecular Materials Group
- Delft University of Technology Kavli Institute of Nanoscience Delft. The Kavli Institute of Nanoscience at Delft University of Technology (The Netherlands) consists of six research groups and a nanofacility cleanroom, all aimed at innovative research at the current frontier of science on a nano-scale.

- Delft University of Technology and Leiden University MSc NanoScience programme. Leiden University and the Delft University of Technology (The Netherlands) offer a unique MSc program on NanoScience, aimed at students who are eager to transcend the traditional borders between scientific disciplines.
- Delft University of Technology Quantum Transport—one of the research groups in the department of Applied Physics
- Delft University of Technology Molecular Biophysics Group
- Drexel Nano Materials Group (Gogotsi's Group)
- Duke University Eom Research Group, Thin Films Laboratory
- Duke University Nanoscience Group
- Earnest — Alumni Magazine for CES at Clemson Nanotechnology research inspires industry collaboration
- Ecole Polytechnique Federale de Lausanne Nanomechanics and Tribology
- Eidgenössiche Technische Hochschule (ETH) NANO-PHYSICS group of the laboratory for solid state physics at ETH Zürich
- Emory University Nie Group. Focus on Drug Delivery, Cancer Therapeutics, Molecular Imaging, Molecular Profiling, and In-vitro Diagnostics.
- Florida State University Molecular Expressions Website
- Free University Berlin Physics Department — Atomic Manipulation Site
- George Washington University Wagner Lab. Synthesis, study and application of solid-state inorganic materials with technologically significant magnetic, electrical, optical, electrochemical or catalytic properties.
- Georgia Institute of Technology Microelectronics Research Center (MiRC). Part of the National Nanotechnology Infrastructure Network (NNIN)
- Georgia Institute of Technology Applied Sensors Laboratory
- Georgia Institute of Technology Zhong Lin (ZL) Wang's Nanoscience and Nanotechnology Group
- Georgia Institute of Technology Nanotechnology and Microfluidics
- Georgia Institute of Technology Nanostructure Research Laboratory
- Georgia Institute of Technology Nanostructure Optoelectronics Group
- Georgia Institute of Technology Molecular Design Institute
- Georgia Tech Research Institute GTRI Nanotechnology Laboratory
- Harvard University Nanoscale Science and Engineering Center—Science of Nanoscale Systems and their Device Applications The Nanoscale Science

and Engineering Center (NSEC) is a collaboration among Harvard University, the Massachusetts Institute of Technology, the University of California at Santa Barbara and the Museum of Science in Boston with participation by Delft University of Technology (Netherlands), the University of Tokyo (Japan), and Brookhaven National Laboratory, Oak Ridge National Laboratory and Sandia National Laboratory. NSF Funded "Nanoscale Science and Engineering Centers"

- Harvard University Center for Imaging and Mesoscale Systems (CIMS). Part of the National Nanotechnology Infrastructure Network (NNIN)
- Harvard University Materials Science Group
- Harvard University Leiber Group—a Physical Chemistry research group in the Department of Chemistry and Chemical Biology of Harvard University
- Harvard University Hongkun Park—physics and chemistry of nanostructured materials
- Harvard University Westervelt group. Department of Physics and Division of Engineering and Applied Sciences
- Harvard University George M. Whitesides Research Group
- Harvard University Nanopore Group
- Harvard University Center for Molecular Imaging Research
- Harvard University Narayanamurti Group
- Haverford College "...to bring together faculty from biology, chemistry, physics, and math in the pursuit of an interdisciplinary effort to develop new protein-based biomaterials for the field of nanometer-scale electronic and mechanical device fabrication."
- Hebrew University of Jerusalem Center for Nanoscience and Nanotechnology
- Helsinki University of Technology Optics and Molecular Materials
- Helsinki University of Technology Materials Physics Laboratory
- Heriot-Watt University, Edinburgh, UK The Nano-Optics Group
- Hong Kong University of Science and Technology Institute of NanoMaterials and NanoTechnology (INMT)
- Howard University Materials Science Research Center. Part of the National Nanotechnology Infrastructure Network (NNIN)
- Illinois Institute of Technology Center on Nanotechnology and Society (CONAS) "... to catalyze informed, inter-disciplinary research and education on the implications of nanoscale science and technology for ethical, legal,

policy, business, and wider social issues, and with a special focus on the human condition."

- Instituto de Carboquímica Group of Carbon Nanostructures and Nanotechnology
- Institutionen för Fysik och Mätteknik (IFM) Department of Physics and Measurement Technology, Biology and Chemistry—Materials Science Division
- Indiana University Advanced Research and Technology Institute
- Institut Für Neue Materialien, University of the Saarland in Saarbrücken (INM) (Germany) Institute for New Materials
- Institute for Solid State and Materials Research, Dresden Germany Materials Science Division
- Interuniversity Microelectronics Center (IMEC) IMEC today is the largest independent microelectronics research and development center in Europe.
- Iowa State University Condensed Matter Physics — Ames Lab
- Iowa State University Condensed Matter Physics — Canfield Research Group
- Iowa State University Surface Engineering and Molecular Assemblies (SEMA) Laboratory
- Institute for Physical High Technology Jena Molecular Nanotechnology Group
- Jawaharlal Nehru Centre for Advanced Scientific Research Carbon Lab. Prof C. N. R. Rao (FRS) Director.
- Joanneum Research—Institute of Nanostructured Materials and Photonics "Our 15 research units make Joanneum Research one of the largest non-university research institutions in Austria." Affliated with international institutions such as the University of California at Berkeley and the Max Planck Institute for Polymer Research in Germany.
- Johns Hopkins University Materials Research Science and Engineering Center (MRSEC). One of 26 MRSECs funded by the National Science Foundation, is composed of scientists at JHU, Brown University, and the National Institute of Standards and Technology (NIST). Research in the Center focuses on nanostructures made from novel materials that exhibit enhanced magneto-electronic properties.
- Johns Hopkins University Dept of Mechanical Engineering
- Kansas State University Dept of Physics, Interdisciplinary Research

- Katholieke Universiteit Leuven Erasmus Mundus Master of Nanoscience and Nanotechnology
- Kaunas University of Technology, LITHUANIA Research Center for Microsystems and Nanotechnology
- Kobe University, Japan Molecular Electronics and Photonics
- Korea Advanced Institute of Science and Technology (KAIST) Superlattice Nanomaterials Laboratory
- Korea University Spintronic Materials Lab., Division of Materials Science and Engineering
- Kyushu University, Japan Division of Nanoelectronics, Dept. of Electronic Device Engineering
- Lehigh University Center for Advanced Materials and Nanotechnology.
- Lehigh University Department of Materials Science and Engineering.
- London Centre for Nanotechnology (LCN) "... designed to act as a focus for current interdisciplinary nanoscale materials and device research." A joint enterprise between University College London and Imperial College.
- Louisiana Tech University Nanosystems Engineering
- Loyola College—Nanobase International Technology Research Institute.
- Ludwig-Maximilians-Universität (LMU), Germany Center for NanoScience (CeNS)
- Ludwig-Maximilians-Universität (LMU), Germany Nanophysics Group
- Lund University The Nanometer Consortium
- Mahidol University Nanotechnology Center. "Building Capability on Molecular Engineering and Devices."
- Marshall University Norton Imaging Group.
- Massachusetts Institute of Technology (MIT)—BioInstrumentation Laboratory Technologies that are currently being developed in the BI Lab include a servo-controlled laser micro-surgical robot, artificial muscle technology based upon conducting polymer biomimetic materials, the Living Chip, a cell-based high throughput screening system for mass-scale discovery of new drugs, and the Nanowalker, a nano-stepping autonomous robotic instrumentation platform.
- FAQ for MIT Institute for Soldier Nanotechnology (ISN)
- MIT Microsystems Technology Laboratories
- MIT Laboratory for Experimental & Computational Micromechanics (LEXCOM)

- MIT Nanostructures Laboratory
- MIT Surfaces and Structures
- MIT NanoMechanical Technology Laboratory
- MIT Space Nanotechnology Laboratory
- Moungi Bawendi's Group MIT "...quantum mechanical behaviors demonstrated by semiconductor and metallic materials..."
- MIT Center for Materials Science and Engineering
- MIT Department of Materials Science and Engineering
- Max Planck Institute for Biophysical Chemistry Department of NanoBiophotonics
- Max Planck Institute for Solid State Research Synthetic Nanostructures Team
- Metroplex Research Consortium for Electronic Devices and Materials (MRCEDM) combines the research and development capabilities of The University of Texas at Arlington, Southern Methodist University, Texas Christian University and The University of North Texas to assist companies that innovate and manufacture advanced technical components and systems.
- Macdiarmid Institute for Advanced Materials and Nanotechnology A New Zealand research organisation concerned with high quality research and research education in materials science and nanotechnology.
- Michigan State University (MSU) Carbon Nanotubes
- MSU—David Tománek "The Nanotube Site"
- MSU—NSF Center on Low-Cost, High-Speed Polymer Composites Processing
- MSU—Composite Materials and Structures Center
- MSU—Advanced Materials Engineering Experiment Station (AMEES) is managed and directed as part of the Composite Materials and Structures Center
- Michigan Technological University Nanotechnology at Michigan Technological University
- McGill — NanoScience & Scanning Probe Microscopy... to apply the tools, techniques and materials developed in the past few years to the emerging new field of Nanoelectronics.
- McGill — NANOSCIENCE AND NANOTECHNOLOGY Course description
- Nagoya University Department of Molecular Design and Engineering
- Nanobiotechnology Center (NBTC) The Nanobiotechnology Center was

established in January 2000 as a Science & Technology Center, with core funding from the National Science Foundation. Nanobiotechnology is an emerging area of scientific and technological opportunity that integrates nano/microfabrication and biosystems to the benefit of both. The Nanobiotechnology Center is characterized by its highly interdisciplinary nature and features a close collaboration between life scientists, physical scientists, and engineers. It has a fully integrated education and outreach effort in which all NBTC faculty participate. The Center brings together experts in their fields from Cornell University, the Wadsworth Center (New York State Health Department in Albany), Princeton University, Oregon Health & Science University, Clark Atlanta University, and Howard University. It also involves the active collaboration of K-12 educators, the Sciencenter Museum in Ithaca, NY, and representatives from industry and the government.

- Nanotechnology Institute (NTI) A collaboration led by Ben Franklin Technology Partners of Southeastern Pennsylvania, Drexel University and The University of Pennsylvania. "... to focus on the transfer of discoveries and intellectual knowledge in the area of nanotechnology from universities to industry partners and on the rapid application and commercialization of this technology to stimulate economic growth."
- Nanyang Technological University Precision Engineering & Nanotechnology Centre
- Nanyang Technological University Micro Electro Mechanical Systems Research
- Nanyang Technological University Advanced Material Research Centre
- Nanyang Technological University Microelectronics Centre
- Nara Institute of Science and Technology Graduate School of Material Science
- National Institute of Advanced Industrial Science and Technology (AIST) Nanomaterials Theory Group
- National Microelectronics Research Centre (NMRC) Ireland Nanotechnology Research
- National Taiwan University Center for Nano Science and Technology (NTU-CNST). "... integrated research projects of this program include nanomaterials, nano-devices, NEMS, and Bio-nano."
- National University of Singapore
- New Jersey Institute of Technology Device & Material Characterization Laboratory

- New Mexico State University Nano Physics
- New York University Center for Advanced Materials and Nanotechnology
- North Carolina Center for Nanoscale Materials (NCCNM) "...research activities in the center are directed towards understanding the fundamental science of nanoscale materials and utilizing their unique properties for commerical applications."
- North Carolina State University Triangle National Lithography Center. Part of the National Nanotechnology Infrastructure Network (NNIN) (a collaboration with the University of North Carolina at Chapel Hill)
- North Carolina State University Analytical Instrumentation Facility
- North Carolina State University Nanoindentation—studying mechanical properties on the atomic scale.
- North Carolina State University Nanoscale Tribology Laboratory.
- North Dakota State University Center for Nanoscale Science and Engineering
- Northeastern University NSF Center for High-rate Nanomanufacturing (CHN)
- Northwestern University Nanoscale Science and Engineering Center (NSEC) for Integrated Nanopatterning and Detection Technologies NSF Funded "Nanoscale Science and Engineering Centers"
- Northwestern University Mirkin Group
- Northwestern University — Institute for Nanotechnology Established as an umbrella organization for the multimillion dollar nanotechnology research efforts at Northwestern University.
- Northwestern University Center for Transportation Nanotechnology.
- Northwestern University MEMS / Nanotechnology.
- Northwestern University Stupp Laboratory
- Northwestern University Nanomechanics
- Northwestern University Mechanics of MEMS and Nanosystems
- Northwestern University Ruoff Group—Nanoscience and Technology Lab
- Northwestern University Center for Nanofabrication and Molecular Self-Assembly
- Northwestern University Professor Horacio Dante Espinosa—Micro and Nanomechanics Lab
- Northwestern University Center for Quantum Devices
- Northumbria University Advanced Materials Research Institute (AMRI).

The Institute's technical focus is in the fields of advanced materials, surface engineering, corrosion and wear.

- Norwegian Institute of Technology (NTH) NMNL (Norwegian Micro and Nano Laboratories), the research infrastructure in Norway hosted by SINTEF (The Foundation for Scientific and Industrial Research at the Norwegian Institute of Technology), the Norwegian University of Science and Technology (NTNU), and the University of Oslo (UiO).
- Norwegian Institute of Technology (NTH) NANOCAPS (at SINTEF).
- Norwegian Institute of Technology (NTH) NANOMAT—Materials Science and Nanotechnology at the Norwegian Microtechnology Center (NMC-MRL) (at SINTEF).
- Nottingham University Nanoscience Group Nanometre scale structures and nanostructured materials play an increasingly pervasive role in a wide range of scientific disciplines ranging from solid state physics through to molecular biology.
- Nottingham Trent University Polymer Engineering Centre
- Oak Ridge National Laboratory (ORNL) Polymer Science Group Summary of ongoing research aimed at simulating nanomachines and aspects of fundamental chemical physics at the molecular scale.
- Ohio University Research Institute to Advance Studies in Nanotechnology
- Ohio University Nanoscale & Quantum Phenomena Institute
- Ohio University Nanospintronics and Nanomagnetics
- Ohio Supercomputer Center Nanotechnology at OSC
- Oklahoma State University NanoNet Probe Microscopy Laboratory
- Oklahoma State University Department of Physics
- Oregon State University College of Engineering
- Oregon State University Microtechnology-Based Energy, Chemical and Biological Systems (MECS)
- Osaka University Protonic NanoMachine Project.
- Oxford University Department of Materials
- Oxford University Begbroke Science Park Institute of Nanotechnology
- Paul Scherrer Institut Laboratory for Micro- and Nanotechnology
- Penn State Atomic-Scale Measurements and Control
- Penn State Center for Molecular Nanofabrication and Devices

- Penn State Nanofabrication Facility. Part of the National Nanotechnology Infrastructure Network (NNIN)
- Penn State Materials Research Institute
- Penn State Allara Research Group
- Penn State Department of Materials Science and Engineering
- PNNL/UW Joint Institute for Nanoscience
- Portland State University Center for Nanoscience and Nanotechnology
- Princeton University Nanostructure Laboratory.
- Princeton University Ceramic Materials Laboratory.
- Purdue Nanotechnology Initiative "...to demonstrate novel techniques for the design and fabrication of nanoelectronic devices by the chemical manipulation of nanometer (10-9 meters) sized clusters and molecular wires."
- Nanotechnology At Purdue (PDF) A visual guide to the various programs and collaborations at Purdue. "This brochure highlights many of those projects across the categories of nanomaterials, nanodevices, nano/bio interfaces, nanomanufacturing, computational nanotechnology, and nanometrology/characterization."
- Purdue University Discovery Park—Birck Nanotechnology Center & Ultra-Performance Nanotechnology Center (ULTEC).
- Purdue University David Janes Group.
- Purdue University Laboratory for Chemical Nanotechnology (PLCN).
- Purdue University nanoHub The nanoHub is a new initiative designed to promote the application of computational science to nanotechnology. Using PUNCH, a network-computing software infrastructure, researchers and educators can access and operate simulation tools through standard Web browsers. Computational scientists can share tools with colleagues, experimentalists can operate simulations tools without downloading and installing them, and users can access supercomputing resources transparently. The focus is on state-of-the-art, research-grade tools designed to complement commercial simulation packages.
- Purdue University Center for Nanoscale Devices.
- Purdue University Nanoscale Physics.
- Purdue University Cluster-Based Materials Group.
- Queen Mary University of London Molecular and Materials Physics Group.

- Research Institute for Technical Physics and Materials Science Budapest, Hungary.
- Rensselaer Polytechnic Institute Nanotechnology Center.
- Rensselaer Polytechnic Institute Center for Directed Assembly of Nanostructures NSF Funded "Nanoscale Science and Engineering Centers"
- Rensselaer Polytechnic Institute Prof. Ajayan's Carbon Nanomaterials Technology Group.
- Rice University Center for Nanoscale Science and Technology
- Rice University Colvin Group
- Rice University Halas Nanophotonics Group
- Rice University Tour Group
- Rice University Laboratory for Nanophotonics
- Rice University Center for Biological and Environmental Nanotechnology NSF Funded "Nanoscale Science and Engineering Centers"
- Rice University Environmental & Energy Systems Insitute of Rice University (EESI)
- RMIT Royal Melbourne Institute of Technology—Nanotechnology Program
- Rochester Institute of Technology NanoPower Research Labs
- Rochester Institute of Technology Nano- and Micro-Electromechanical Systems, Nano- and Microsystems, Nanoarchitectronics
- Rutgers University Bio Nano Robotics
- Rutgers University Division of Computer & Information Sciences
- Rutgers University Laboratory for Surface Modification (LSM)
- Seoul National University Center for Science in Nanometer Scale
- Sheffield Hallam University Electronics Research Group, Microsystems & Machine Vision Laboratory (MMVL)
- South Dakota School of Mines and Technology Center for Accelerated Applications at the Nanoscale (CAAN).
- Stanford University Nanofabrication Facility—serves academic, industrial, and governmental researchers across the US in areas ranging from of optics, MEMS, biology, and chemistry, as well as traditional electronics device fabrication and process characterization. Part of the National Nanotechnology Infrastructure Network (NNIN)
- Nanoscience and Nanotechnology Online Seminars

- Explore the universe of nanotubes and bucky balls. Expand your vision of a changed world, from the quality of our goods to the quality of our lives. Presenting the latest nanoscience and nanotechnology concepts, the Stanford Engineering and Science Institute will explore the promise of a wide range of exciting new products and applications capable of transforming and redefining many industries. Learn from Stanford faculty and industry experts the potentially broad impacts of nanotechnology for your business. Learn More
- Stanford University Department of Materials Science and Engineering
- Stanford University Center for Molecular and Genetic Medicine
- Stanford University Geballe Laboratory for Advanced Materials (LAM)
- Stanford University Chemical Engineering
- Stanford University Mechanical Engineering
- Stanford University S.C.Zhang Group
- http://buckminster.physics.sunysb.edu/ Department of Physics and Astronomy.... study the fundamental properties of Fullerenes (Carbon-60) and doped fullerenes
- Stockholm University Sol-Gel Group at the Department of Inorganic Chemistry
- http://www.matscieng.sunysb.edu/tsl/nano/ Thermal Spray Lab
- SUNY—University at Buffalo The Center for Advanced Photonic and Electronic Materials (CAPEM)—Laboratory for Spintronics Research in Semiconductors (LSRS)
- SUNY—University at Buffalo Electronic Packaging Laboratory
- SUNY — Molecular Structure Laboratory Department of Chemistry at SUNY Stony Brook.
- Swiss Federal Institute of Technology Zurich (ETH Zurich) FIRST! Center for Micro- and Nanoscience (Frontiers in Research: Space and Time)
- Tata Institute of Fundamental Research (TIFR) Department of Condensed Matter Physics and Materials Science
- Technical University of Denmark NANO@DTU—organizes the broad spectrum of nanotechnology research and education at DTU.
- Technical University of Denmark Nanotechnology Research Group.
- Technical University of Denmark Center for Atomic-scale Materials Physics.
- Technion—Israel Institute of Technology Microelectronics Research Center

- Technische Universitat Berlin Project Nanostructures
- Tel Aviv University TAU Research Institute for Nanoscience and Nanotechnology
- Texas Christian University Atom Optics Lab
- Texas Engineering Experiment Station (TEES) a member of Texas A&M University System
- Texas A&M University Texas Institute of Intelligent Bio-Nano Materials and Structures for Aerospace Vehicles (TiiMS)
- Texas State—Nanomaterials Applications Center—NAC
- Texas Nanotechnology Initiative A Consortium of industry, universities, government, and venture capitalists whose goal is to position Texas as the Nanotechnology State
- Towson University Nanotechnology Studies using Scanning Probe Microscopy
- Towson University Nanotechnology Laboratory
- Trinity College Dublin Nanomag Research Group
- Trinity College Dublin Centre for Research on Adaptive Nanostructures and Nanodevices (CRANN)
- TRI/Princeton "... TRI has extended its expertise to provide advanced research and education in polymers, fibers, films, human hair, and porous materials."
- Tufts University The Walt Group... nanosensors and nanostructures.
- University at Albany—SUNY Albany NanoTech "is a fully-integrated research, development, prototyping, pilot manufacturing and education resource managing a strategic portfolio of state-of-the-art laboratories, supercomputer and shared-user facilities and an array of research centers located at the University at Albany—SUNY." Formerly "The Center for Advanced Thin Film Technology".
- University at Albany College of Nanoscale Science and Engineering. "... cross-disciplinary curriculum integrates the fundamental science principles of physics, chemistry, computer science, and biology with the cross cutting fields of nanosciences, nanoengineering, nanoeconomics, and nanobiology."
- University at Albany—SUNY Advanced Computer Modeling for Nanosystems and Processes.
- University at Albany—SUNY Center for Advanced Thin Film Technology—Microsystems Integration Laboratory

- Université Catholique de Louvain UCL/CRMN—Centre de Recherche en Dispostifs et Matériaux Electroniques Micro- et Nanoscopiques (CeRMiN). Research Center in Micro and Nanoscopic Materials and Electronic Devices
- University College Cork Nanotechnology Group—Dr. Gareth Redmond Director. "The Group undertakes research aimed at development and innovative exploitation of nanoscale materials and devices within emerging Information and Communication Technology (ICT) application areas."
- University of Aalborg Institute of Physics and Nanotechnology
- University of Aarhus and Aalborg University iNANO—interdisciplinary Nanoscience Center
- University of Akron Center for Molecular Design and Recognition
- University of Alberta Faculty—National Institute for Nanotechnology "... an integrated, multi-disciplinary institution involving researchers in physics, chemistry, engineering, biology, informatics, pharmacy and medicine. Established in 2001, it is operated as a partnership between the National Research Council and the University of Alberta, and is jointly funded by the Government of Canada, the Government of Alberta and the university."
- University of Alberta Buriak Group
- University of Alberta Centre for Nanoscale Physics
- University of Alberta Microsystems Technology Research Institute—NanoFab. "... an open access micro and nano fabrication facility for use by the University of Alberta research community and other groups including Canadian universities and companies involved in microsystem research and related prototyping development."
- University of Arizona Nanomechanics and Mesoscopic Physics
- University of Arizona AFM Lab
- University of Arkansas Semiconductor Fabrication and NanoScale Characterization Facility
- University of Basel National Center of Competence in Research Nanoscale Science
- University of Basel Nano-Electronics (Mesoscopic Physics)
- University of Bath, UK Materials Research Centre
- University of Bath, UK Centre for Electron Optical Studies
- University of Berlin — Project Nanostructures Focuses on the fabrication, characterization and theoretical modelling of quantum wires and quantum dots.

- University of Birmingham I^2 NanoTech Centre
- University of Birmingham Nanoscale Chemistry
- University of Birmingham Nanoscale Physics Research Laboratory
- University of Birmingham μ-Nano Consortium
- University of California at Berkeley • Davis • Merced • Santa Cruz The Center for Information Technology Research in the Interest of Society (CITRIS). Includes as members/participants California High-Tech Industry & The State of California.
- University of California at Berkeley Stacy Group "... involved in the development of new synthetic routes that lead to the discovery of materials for emerging technologies."
- University of California at Berkeley Berkeley Sensor & Actuator Center—MEMS Reference Database
- University of California at Berkeley Microstructured Materials Group
- University of California at Berkeley Professor Connie Chang-Hasnain Optoelectronics Research Group
- University of California at Berkeley McEuen Group has moved to McEuen Group—LASSP Cornell University—Laboratory of Atomic and Solid State Physics—Department of Physics
- University of California at Berkeley Alivisatos Group
- University of California at Berkeley Zettl Research Group—Condensed Matter Physics—Department of Physics. An experimental solid state physics group.
- University of California at Berkeley, UC San Francisco, UC Santa Cruz Institute for Bioengineering, Biotechnology and Quantitative Biomedical Research (QB3). QB3 is one of four institutes established through the California Institutes for Science and Innovation
- University of California at Davis, NEAT Nanophases in the Environment, Agriculture, and Technology is a multidisciplinary research and education program which links the fundamental physics, chemistry, and engineering of small particles and nanomaterials to several challenging areas of investigation.
- University of California, Irvine Integrated Nanosystems Research Facility (INRF)
- University of California at Los Angeles Heath Group—Metal & Semiconductor nanocrystal research.

- University of California at Los Angeles Nanoelectronics Research Facility
- University of California at Los Angeles Optoelectronics Group, headed by Prof. Eli Yablonovitch.
- University of California at Los Angeles Stoddart Research Group
- University of California at Los Angeles Wudl Group
- University of California, Riverside Center for Nanoscale Science and Engineering
- University of California, Riverside Nanotechnology Educational Outreach
- University of California, Riverside Biomedical Science & Nanotechnology Laboratory
- University of California at Los Angeles and University of California at Santa Barbara California NanoSystems Institute (CNSI). UCLA & UCSB have joined to build the California Nanosystems Institute, which will facilitate a multidisciplinary approach to develop the information, biomedical, and manufacturing technologies that will dominate science and the economy in the 21st century. CNSI is one of four institutes established through the California Institutes for Science and Innovation
- University of California at Santa Barbara Nanofabrication Facility. Part of the National Nanotechnology Infrastructure Network (NNIN)
- University of California at Santa Barbara Nanoscale Physics—Cleland Group
- University of California at Santa Barbara Awschalom Group
- University of California at Santa Barbara Institute for Quantum Engineering, Science and Technology (iQUEST) (formerly known as the Quantum Institute)
- University of California at Santa Barbara Department of Molecular, Cellular and Developmental Biology (MCDB)
- University of California Santa Cruz Quantum Electronic Group
- UCSC Extension General Interest Page (Current nanotechnology course offerings)
- University of California at San Diego Physics Department—Thin Film Lab—Nanoscience Group. Professor Ivan K. Schuller Director.
- University of California at San Diego Materials and Devices Layer at UCSD. Professor Ivan K. Schuller Director.
- University of California at San Diego MURI Integrated Nanostructured Supersensors. Professor Ivan K. Schuller Director.

- University of California at San Diego Silva Research Group
- University of California at San Diego and UC Irvine California Institute for Telecommunications and Information Technology (Cal-(IT)2). Cal-(IT)^2 is one of four institutes established through the California Institutes for Science and Innovation initiative proposed in the year 2000 by Governor Gray Davis. Cal-(IT)^2, a partnership between UC San Diego and UC Irvine, seeks to ensure that California maintains its leadership in the telecommunications and information technology marketplace.
- University of Cambridge CUED Nanoscale Science Laboratory
- University of Cambridge Centre of Molecular Materials for Photonics and Electronics
- University of Cambridge the Cavendish Laboratory is the Department of Physics
- University of Cambridge The Nanoscience Centre
- University of Cambridge IRC in Nanotechnology—The Interdisciplinary Research Collaboration (IRC) in Nanotechnology is funded by EPSRC, BBSRC, MRC and the MoD as a collaboration between the University of Cambridge, University College London and the University of Bristol.
- University of Canterbury—NEST Nanostructure Engineering Science and Technology Group
- University of Canterbury Cluster Assembled Nanostructures—Dr. Simon Brown, Research Interests
- Université Catholique de Louvain Research Center on Micro and Nanoscopic Electronic Devices and Materials (CERMIN)
- University of Chicago Materials Research Science and Engineering Center
- University of Cincinnati Nanoelectronics Laboratory
- University of Colorado Team Weimer
- University of Connecticut Nanostructured Materials Program
- University of Copenhagen, Denmark Center for Interdisciplinary Studies of Molecular Interactions.
- University of Copenhagen, Denmark Nano-Science Center
- University of Delaware Center for Molecular and Engineering Thermodynamics (CMET)
- University of Delaware Center for Nanomachined Surfaces
- University of Dublin Trinity College Centre for Research on Adaptive Nanostructures and Nanodevices.

- University of Durham Department of Physics.
- University of Durham Carbon Nanostructures.
- University of Durham Durham Nanoscale Magnetics Group.
- University of Durham Nanotechnology University Innovation Centre (Nanotechnology UIC).
- University of Glasgow Nanoelectronics Research Centre. Provides a focus for diverse research activities within the University of Glasgow linked by a common interest in Nanoelectronics and Nanofabrication.
- University of Glasgow Spectroscopy of Nanostructures
- University of Glasgow Kelvin Nanotechnology
- University of Greenwich NANO-SCIENCE SIMULATION GROUP
- University of Hamburg Nanostructure Physics Group
- University of Hannover, Germany Institute for Solid State Physics—Department of Nanostructures
- University of Houston Texas Center for Superconductivity (and Advanced Materials) at the University of Houston (TCSUH)
- University of Houston Sharma Group—Mechanical Engineering.
- University of Illinois at Chicago Materials Characterization Laboratory
- University of Illinois at Chicago Microfabrication Applications Laboratory (MAL)
- University of Illinois at Chicago Nanoscale structures of asphaltene steric-colloid and asphaltene micelles & vesicles.
- University of Illinois at Urbana-Champaign Center for Nanoscale Science & Technology (CNST)
- University of Illinois at Urbana-Champaign Beckman Institute for Advanced Science and Technology
- University of Illinois at Urbana-Champaign Beckman Institute for Advanced Science and Technology—Molecular and Electronic Nanostructures
- University of Illinois at Urbana-Champaign Micro and Nanotechnology Laboratory
- University of Illinois at Urbana-Champaign Frederick Seitz Materials Research Laboratory (MRL).
- University of Illinois at Urbana-Champaign MASS Group. "...research in all aspects of the emerging technology of Microelectromechanical Systems."
- University of Jyväskylä Nanoscience Center

- University of Kentucky Materials Research Science and Engineering Center (MRSEC) Advanced Carbon Materials Center
- University of Leeds Centre for Self-Organising Molecular Systems
- University of Leeds Centre for Nano-Device Modelling
- University of Leeds Masters Training Package (MTP) in Nanoscale Science and Technology, run jointly by the universities of Leeds and Sheffield.
- University of Leipzig (Universität Leipzig) Semiconductor Physics Group. Quantum dot physics and device applications, and hybrid organic-anorganic nanostructure research.
- University of Liverpool Centre for Nanoscale Science "... material aspects of nanotechnology and nanoparticle research."
- University of Liverpool Chemistry with Nanotechnology—MChem.
- University of London Nano4CE "... a specialist unit operated by the University with the remit of working with industry to develop product forms that utilise nanomaterials."
- University of Louisville Nanoscale Imaging Facility
- University of Louisville Chemical Engineering Department
- University of Louisville Electro-Optics Research Institute
- University of Louisville MicroTechnology Center
- University of Manchester Manchester Centre for Mesoscience & Nanotechnology
- University of Manchester Nanotechnology Research
- University of Maryland Maryland Center for Integrated Nano Science and Engineering (M-CINSE)
- University of Maryland, College Park — Robert H. Smith School of Business Asian Technology Information Program.
- University of Maryland Materials Research Science and Engineering Center (MRSEC)
- University of Maryland Fuhrer Research Group / Nanoelectronics Research Group
- University of Maryland Drew Research Group / Nano-optics
- University of Maryland LPS Nanostructures Team
- University of Maryland The Williams Lab / Surface Physics
- University of Maryland Co-laboratory for Nanoparticle Based Manufacturing & Metrology—NM^2

- University of Massachusetts Amherst MassNanoTech
- University of Massachusetts Amherst Dinsmore Research Group.
- University of Massachusetts Amherst Materials Research Science and Engineering Center (MRSEC).
- University of Massachusetts Amherst Polymer Science & Engineering Department
- University of Massachusetts Lowell Nano Manufacturing.
- University of Michigan Solid State Electronics Laboratory. Part of the National Nanotechnology Infrastructure Network (NNIN)
- University of Michigan Center for Biologic Nanotechnology
- University of Michigan WIMS Engineering Research Center (ERC) an NSF-University-Industry Collaboration
- University of Michigan Glotzer Group—Laboratory for Computational Nanoscience and Soft Matter Simulation
- University of Michigan Materials Science & Engineering
- University of Michigan Kotov Group—Nano Research Laboratory
- University of Minnesota Minnesota Nano Technology Cluster. Part of the National Nanotechnology Infrastructure Network (NNIN)
- University of Minnesota Center for NanoEnergetics Research (CNER). CNER is an Army funded center created in the spring of 2001 and exists at four university sites, with the University of Minnesota as the lead institution.
- University of Minnesota Nanotechnology Coordinating Office
- University of Minnesota Particle Technology Laboratory (PTL)
- University of Minnesota Molecular Nanosciences Alliance for Interdisciplinary Studies and Activities (MONALISA)
- University of Minnesota Norris Research Group
- University of Montreal Pavillon J.-Armand Bombardier: Nanosciences and Nanotechnologies
- University of Muenster Interface Physics Group
- University of Nebraska—Lincoln Center for Materials Research and Analysis (CMRA)
- University of Nebraska—Lincoln Nanomachining/Nanofabrication at the Center for Electro-Optics.

- University of Nebraska—Lincoln Quantum Device Laboratory
- University of Nebraska—Lincoln Condensed Matter/Materials Physics and Nanotechnology.
- University of Nebraska Medical Center Nanomedicine Group
- University of Nevada, Reno Adams Group
- University of Newcastle upon Tyne Centre for Nanoscale Science. A Multidisciplinary research centre linking scientists, engineers, medical researchers, and clinicians developing micro- and nanodevices and technologies for applications in biology, biotechnology and medicine.
- University of Newcastle upon Tyne Institute for Nanoscale Science and Technology (INSAT)
- University of Newcastle upon Tyne INEX—a microsystems and nanotechnology R&D, commercialisation and manufacturing organisation.
- University of New Mexico Integrative Nanoscience and Microsystems
- University of New Hampshire The Nano Group
- University of New Hampshire Nanotechnology and Society Research Group
- University of New Hampshire Nanostructured Polymers Research Center
- University of New Hampshire Nanostructured Polymers Research Center—Polymer Research Group (PRG).
- University of New Hampshire Nanostructured Polymers Research Center—Polymer Nanoparticle Laboratory (PNL).
- University of New Hampshire Nanostructured Polymers Research Center—Advanced Polymer Laboratory (APL).
- University of New Mexico Nanomaterials and Nanocharacterization. Part of the National Nanotechnology Infrastructure Network (NNIN)
- University of New Mexico Brinker Group.
- University of North Carolina at Chapel Hill North Carolina Center for Nanoscale Materials (NCCNM)
- University of North Carolina at Chapel Hill Nanoscale Science Research Group (NSRG)
- University of North Carolina at Chapel Hill The nanoManipulator
- University of Notre Dame Center for Nano Science and Technology
- University of Notre Dame Nanofabrication Facility
- University of Notre Dame Nano Science and Technology Center

- University of Nottingham Nottingham Nanotechnology Centre (includes 6 specializations: Quantum Nanostructures, Synthesis of Nanoscale Vessels and Storage Devices, Single Molecule Nanoscience, Biofunctional Surfaces and Materials, Nanofabrication and Characterisation, Biological Application of Nanoparticles)
- University of Oregon Jim Hutchison Lab
- University of Oregon Materials Science Institute
- University of Oregon CAMCOR—Center for Advanced Materials Characterization in Oregon. "... a full-service, comprehensive materials characterization center available to research institutions and private industry."
- University of Oregon Oregon Center for Optics
- University of Oxford Carbon and Nanotechnology Group
- University of Oxford Carbon Nanotube Group
- University of Oxford Nanotechnology Group Pages
- University of Oxford Centre for Quantum Computation
- University of Pennsylvania LRSM / Laboratory for Research on the Structure of Matter
- University of Pennsylvania School of Engineering and Applied Science
- University of Pennsylvania Nano/Bio Interface Center
- University of Pittsburgh Swanson Center for Micro and Nano Systems
- University of Queensland (UQ) Nano Optics Group.
- University of Queensland Centre for Microscopy and Microanalysis.
- University of Queensland Soft Condensed Matter Group.
- University of Queensland NanoMaterials Centre.
- University of Queensland Nanotechnology & Biomaterials Centre.
- University of Queensland ARC Centre for Functional Nanomaterials (In partnership with the University of New South Wales, the Australian National University and the University of Western Sydney).
- University of Rochester Institute of Optics—Nano-Optics Group
- University of Sheffield Masters Training Package (MTP) in Nanoscale Science and Technology, run jointly by the universities of Leeds and Sheffield.
- University of Sheffield Nanorobotics Technologies for Multidimensional Imaging and Manipulation of Nanoobjects

- University of Southern California Laboratory for Molecular Robotics
- University of Southern California The Koel Group—Surface Science: Understanding Reactions at Nanoparticles and Catalysts
- University of Southern California Nanotechnology & Micro Electro-Mechanical Systems Research Area
- University of South Carolina NanoCenter
- University of South Carolina USC NIRT: Philosophical and Social Dimensions of Nanoscale Research
- University of Southhampton Nanoscale Systems Integration Group (NSI).
- University of Sussex Fullerene Research Centre.
- University of Sussex Nanoscience and Nanotechnology @ Sussex.
- University of Sydney Nanostructural Analysis Network Organisation—Major National Research Facility (NANO—MNRF). Australian facility for nanometric analysis of the structure and chemistry of materials in both physical and biological systems. An unincorporated joint venture between university, industry, state and federal governments.
- University of Technology Sydney Institute for Nanoscale Technology
- University of Technology Sydney Institute for Nanoscale Technology Nanohouse Initiative.
- University of Texas at Arlington NanoFAB Research and Teaching Facility
- University of Texas at Austin Microelectronics Research Center (MRC). Part of the National Nanotechnology Infrastructure Network (NNIN)
- University of Texas at Austin Center for Nano- and Molecular Science and Technology
- University of Texas at Austin Science, Technology and Society (STS) "Explore social impacts of rapid scientific and technological change."
- University of Texas at Dallas NanoTech Institute.
- University of Texas at Dallas Center for Quantum Electronics.
- University of Texas at Dallas MEMS Research Group
- University of Tokyo Quantum Microstructure Devices—Sakaki Laboratory
- University of Toronto Energenius Centre for Advanced Nanotechnology (ECAN)
- University of Toronto Electronic Materials Group
- University of Toronto Geoff A. Ozin Group

- University of Toronto Nanoclub aims to promote nanotechnology at the University of Toronto. We are entirely student-run, but receive support from the Department of Materials Science Engineering.
- University of Twente—Netherlands Strategic Research Orientation Nanolink.
- University of Ulster Nanotechnology Research Institute
- University of Vienna Institute of Materials Physics
- University of Virginia MRSEC—Center for Nanoscopic Materials Design
- University of WalloniaNanoWal—Wallonia Network for Nanotechnologies
- University of Warwick — The Centre for Nanotechnology and Microengineering "...a national centre for excellence in nanotechnology and ultraprecision engineering for both the academic community and manufacturing industries."
- University of Washington Center for Nanotechnology
- University of Washington Center for Nanotechnology—Nanotech Student Association. "NSA fosters interdisciplinary research in nanotechnology, provides a forum for exchange of ideas between students from different disciplines, and promotes nanotechnology at the university level."
- University of Washington NanoTech User Facility (NTUF). Part of the National Nanotechnology Infrastructure Network (NNIN)
- University of Washington Center for Applied Microtechnology (CAM)
- University of Washington Department of Materials Science & Engineering
- University of Washington Molecular Bioengineering & Nanotechnology
- University of Waterloo Nanotechnology Engineering Program
- University of Wisconsin, Madison Nanoscale Science and Engineering Center (NSEC)
- University of Wisconsin, Madison Center for NanoTechnology
- University of Wisconsin, Madison Nanowires
- University of Wisconsin, Madison Applied Superconductivity Center
- University of Wisconsin, Madison Center for Plasma-Aided Manufacturing
- University of Wisconsin, Madison Materials Research Science and Engineering Center on Nanostructured Materials and Interfaces (MRSEC)
- University of Wisconsin, Madison Learning how to Fabricate Nanowires
- University of Wisconsin, Madison Nealey Research Group

- University of Wollongong, Australia Fundamental Properties of Semiconductors Research Centre.
- University of Wollongong, Australia ARC Centre for Nanostructured Electromaterials.
- Uppsala University Division of Materials Science at the Ångström Laboratory.
- University of York York-JEOL Centre: a collaboration between the University of York, Yorkshire Forward and JEOL UK Ltd.
- University of York Nanoscale Quantum Simulations for Nanostructures and Advanced Materials—Nanoquanta
- Vanderbilt University Vanderbilt Institute of Nanoscale Science and Engineering (VINSE).
- Virginia Tech Physics Department—Research Areas—Centers & Institutes.
- Virginia Tech Fiber & Electro-Optics Research Center.
- Virginia Tech Center for Modeling and Simulation in Material Science.
- Wake Forest The Center for Nanotechnology.
- Walter Schottky Institute—Technische Universität München Experimental Semiconductor Physics I (E24)—Prof. Gerhard Abstreiter.
- WASEDA University Establishment of Molecular Nano-Engineering by Utilizing Nanostructure Arrays and its Development into Micro-Systems.
- Washington University in St. Louis Buhro Group.
- Wayne State University Brock Group—Inorganic, Solid State, and Materials Chemistry.
- Wayne State University Jaewu Choi's Group: nano Devices & Systems Laboratory.
- Weizmann Institute of Science Prof. Ehud Shapiro—Laboratory for Biological Nanocomputers & The BIOSPI Project.
- Weizmann Institute of Science Department of Condensed Matter Physics.
- Weizmann Institute of Science Department of Materials and Interfaces.
- Western Michigan University Nanotechnology Computation and Research Center (NRCC) "NRCC's unique niche area is "nanobioenvironmental chemistry" which will build on Western's existing research strengths in nanotechnology approaches to environmental chemistry, biotechnology, and interrelated phenomena and problems."
- Widener University Fullerenes At Widener.

- Albany Institute of Nanotechnology and Applied Sciences—Research facility affiliated with the University at Albany engaged in a variety of nanoscience fields, including MEMS. Pictures and facility descriptions, news, events, staff profiles, bibliography.
- Alberta University—Department of Physics, Centre for Nanoscale Physics: ultrafast microscopy, ultrafast spectroscopy, nanocrystals and ion implantation, modelling of dynamics at solid surfaces, superconductivity, phase transitions and biophysics.
- Analytical Instrumentation Facility—North carolina state university, providing studies in instruments electron & scanned probe microscopy.
- Australian National University—Nanotube group: producing boron nitride and carbon nanotubes using a new relatively low temperature milling process.
- Birmingham University—Nanoscale Physics Research Laboratory: surface modification, cluster physics, sensors, scanning probe microscopy, nano-optics, and thin films.
- California University—Department of Physics: Condensed Matter Physics: Zettl Research Group: research into the electronic, magnetic, and structural properties of novel materials, including nanostructures such as telescoping nanotube bearings. (Berkeley)
- Cambridge University—Department of Engineering: Nanoscale Science Laboratory: nanotubes, nanomagnetics, nanowires, spin dependent transport, nanotribology, sensors, scanning probe microscopy, electron beam lithography, and an introduction to nanotechnology.
- Center for Nano and Molecular Science and Technology at UT Austin—Center for Nano- and Molecular Science and Technology (CNM)is a multidisciplinary, collaborative research center in the University of Texas at Austin. The mission of the CNM is to foster education, science, and engineering in nanoscience and nanotechnology at the University of Texas.
- Cranfield University—Nanotechnology Group, Advanced Materials Department: Research, facilities and courses, including the MSc in Microsystems and Nanotechnology, the first nanotechnology degree course in Europe.
- Georgia Tech—Center for Nanoscience and Nanotechnology. Nanostructure synthesis, sensors, MEMS, photonic crystals, microscopy, nano-lithography.
- Illinois University—The Beckman Institute for Advanced Science and Technology: molecular and electronic nanostructures; computational

electronics; scanning tunneling microscopy, including lithography and fabrication; semiconductor nanostructures and photonics; synthesizing and characterizing new materials.

- Imperial College—Undertakes programme of one-year interdisciplinary nanomaterials research masters course. Provides funds as well.
- Integrated Microfabrication Lab at BYU—Semiconductor process laboratory, used for research in solid state electronics, optoelectronics, and microfabrication..
- Laboratory for Thin Films Physics—Studies size effects in phase transitions, structure and surface phenomena in condensed films of various materials. Includes research interests, people, publications, photo gallery and links.
- Newcastle upon Tyne University—Centre for Nanoscale Science & Technology: multidisciplinary research centre linking scientists, engineers, medical researchers, and clinicians developing micro and nanodevices and technologies for applications in biology, biotechnology and medicine.
- Northwestern University—Transportation Center: nanotechnology as applied to the transportation sector.
- Northwestern University—Nanoscience and Technology Lab: synthesis and properties of carbon and boron nitride nanotubes and nanofibers, fullerenes, and how materials synthesis at the sub-micron scale can impact the biological sciences.
- Nottingham University—Nanoscience Group: "bottom up" approaches to nanotechnology, scanning probe microscopes, properties of fullerenes and derivatives.
- Purdue University—Nanoscale Physics: research into nanotubes, nanoclusters, molecular wires, break junctions, scanning probe microscopy (STM and AFM), and field-ion microscopy.
- Rensselaer Polytechnic Institute—Advanced Materials & Nanotechnology Research
- Texas University—Malcolm Brown's Nanopage: ongoing research in the structure and synthesis of nanostructures and high resolution transmission electron microscopy and specialized light microscopy.
- Washington University—Center for study and research in nanotechnology, offering cross-discipline graduate courses in the field.

Science Technology: Nanotechnology Books

- Engines of Creation: The Coming Era of Nanotechnology—By K. Eric

Drexler; Anchor Books, 1986, ISBN 0385199732. With a Foreword by Marvin Minsky, this is the first and still the classic book on nanotechnology. [Foresight Institute, Online]

- AsiaNANO 2002: Proceedings of the Asian Symposium on Nanotechnology and Nanoscience 2002—Edited by Masatsugu Shimomura, Teruya Ishihara; World Scientific Publishing, 2002, ISBN 9812383921. Conference proceedings, Tokyo, Japan, 27-29 November 2002, table of contents. [publisher website]
- Becoming Immortal: Nanotechnology, You, and the Demise of Death—By Wesley M. Du Charme; Blue Creek Ventures, 1995, ISBN 0964628201. Popular treatment of nanotechnology, cryonics. [Amazon.com]
- Chemistry at the Beginning of the Third Millennium—Edited by Luigi Fabbrizzi, Antonio Poggi; Springer-Verlag, 2000, ISBN 0387989889. Accessible treatment of molecular design, supramolecules, nanotechnology and beyond; 15 chapters by specialists. [publisher website]
- Designing the Molecular World—Descriptions, reviews, purchase source. [Amazon.com]
- Designing the Molecular World: Chemistry at the Frontier—Philip Ball; Princeton University Press, 1996, ISBN 0691029008. Accessible survey of new results in many major fields allied with chemistry: molecular electronics, artificial enzymes, smart polymer gels. [Princeton University Press]
- Electrochemical Nanotechnology: In Situ Local Probe Techniques at Electrochemical Interfaces—Edited by Wolfgang J. Lorenz, Waldfried Plieth; John Wiley & Sons, 1998, ISBN 0471575186. Full treatment of interface phenomena at solid state surfaces by local probe methods such as near contact microscopy. [publisher website]
- Electrochemical Nanotechnology: In Situ Local Probe Techniques at Electrochemical Interfaces—Descriptions, reviews, purchase source. [Amazon.com]
- Engines of Creation: The Coming Era of Nanotechnology—Description, picture, links. [Anchor]
- Engines of Creation: The Coming Era of Nanotechnology—Descriptions, reviews, purchase source. [Amazon.com]
- The First Immortal—By James L. Halperin; Del Rey, Random House, 1998, ISBN 0345420926. Fictionally explores nanotechnology's impact on the human future, special focus on cryonics, life extension. Has author forum, news, information, opinion poll. [Random House]
- The First Immortal—Descriptions, reviews, purchase source. [Amazon.com]

- The Forge of Mars—By Bruce Balfour; ACE-Oryx, 2002, ISBN 0441009549. Futuristic thriller explores nanotechnology, artificial intelligence, colonizing Mars. Site has science news, author bio, reader's guide, links, Mars photos. [Scribbling Gargoyle Entertainment Corp.]
- The Forge of Mars—Descriptions, reviews, purchase source. [Amazon.com]
- Future Trends in Microelectronics: The Nano Millennium—Descriptions, reviews, purchase source. [Amazon.com]
- Future Trends in Microelectronics: The Nano Millennium—Edited by Serge Luryi, Jimmy Xu, Alex Zaslavsky; John Wiley & Sons, 2002, ISBN 0471212474. Based on ideas, discussions arising from 3rd Future Trends in Microelectronics (FTM) workshop series, summer 2001. [publisher website]
- Hacking Matter: Levitating Chairs, Quantum Mirages, and the Infinite Weirdness of Programmable Atoms—By Wil McCarthy; Basic Books, 2003, ISBN 046504428X. Programmable matter, via quantum dots, will someday change our lives. Author visits labs of major firms, talks to researchers developing this extraordinary technology. [SFF Net]
- Hacking Matter: Levitating Chairs, Quantum Mirages, and the Infinite Weirdness of Programmable Atoms—Descriptions, reviews, purchase source. [Amazon.com]
- Handbook of Nanoscience, Engineering, and Technology—Edited by William A. Goddard III, Donald W. Brenner, Sergey Edward Lyshevski, Gerald J. Iafrate; CRC Press, 2002, ISBN 0849312000. Examines NEMS applications in many diverse areas. [CRC Press]
- Handbook of Nanoscience, Engineering, and Technology—Descriptions, reviews, purchase source. [Amazon.com]
- Handbook of Nanostructured Materials and Nanotechnology—Edited by Hari Singh Nalwa; Academic Press, 2000, ISBN 0125137605, Volumes 1-5. Won 1999 Award of Excellence in Engineering Handbooks from Association of American Publishers; 140 authors. [Academic Press]
- Handbook of Nanostructured Materials and Nanotechnology—Descriptions, reviews, purchase source. [Amazon.com]
- Highly-Organized Substances and Nanotechnology: 2nd International Conference on Chemistry of Highly-Organized Substances and Scientific Principles of Nanotechnology—Edited by V.B. Alescovskii; John Wiley & Sons, 1999, ISBN 3527298045. Part of macromolecular symposia series, meeting theme: combining isolated effects in studying covalent and non-covalent multinuclear chemical systems. [publisher website]

- Integrated Chemical Systems: A Chemical Approach to Nanotechnology—By Allen J. Bard; Wiley-Interscience, 1994, ISBN 0471007331. Systematic, real world approach to nanosystems, many actual examples, scores of figures and photo illustrations. [John Wiley & Sons]
- Integrated Chemical Systems: A Chemical Approach to Nanotechnology—Descriptions, reviews, purchase source. [Amazon.com]
- Introduction to Nanotechnology—Descriptions, reviews, purchase source. [Amazon.com]
- Introduction to Nanotechnology—By Charles P. Poole, Jr., Frank J. Owens; John Wiley & Sons, 2003, ISBN 0471079359. Broad, practical overview of topic, gives researchers coverage of developments in nanotechnology outside their fields. [publisher website]
- The Investor's Guide to Nanotechnology & Micromachines—Descriptions, reviews, purchase source. [Amazon.com]
- The Investor's Guide to Nanotechnology & Micromachines—By Glenn Fishbine; John Wiley & Sons, 2002, ISBN 0471443557. First investment guide to fast emerging industry. In 2000, government and private investment topped $1 billion, should grow exponentially in future. [John Wiley & Sons]
- Key Technologies for the 21st Century: Scientific American: A Special Issue—Descriptions, reviews, purchase source. [Amazon.com]
- Key Technologies for the 21st Century: Scientific American: A Special Issue—Edited by Scientific American; W.H. Freeman & Co, 1996, ISBN 0716729482. One of series; treats future of medicine, cyberspace, transport, production, energy, environment, with a nanotechnology focus. [publisher website]
- MEMS and NEMS: Systems, Devices, and Structures—By Sergey Edward Lyshevski; CRC Press, 2002, ISBN 0849312620. Multidisciplinary, forms basis for MEMS, NEMS synthesis, modeling, analysis, simulation, control, prototyping, fabrication. [CRC Press]
- MEMS and NEMS: Systems, Devices, and Structures—Descriptions, reviews, purchase source. [Amazon.com]
- Microelectronics, Microsystems and Nanotechnology—Edited by Androula G. Nassiopoulou, Xanthi Zianni; World Scientific Publishing, 2000, ISBN 9810247699. Conference proceedings, Athens, Greece, 20-22 November 2000, table of contents. [publisher website]
- Microelectronics, Microsystems and Nanotechnology—Information, purchase source. [Amazon.com]

- Molecular Engineering of Nanosystems—Descriptions, reviews, purchase source. [Amazon.com]
- Molecular Engineering of Nanosystems—By Edward A. Rietman; Springer-Verlag NY, 2001, ISBN 0387989889. Technical introduction. It is growing possible to manipulate biotic tissues and materials the size of individual cells, organelles, molecules. [publisher website]
- Nano- and Microelectromechanical Systems: Fundamentals of Nano- and Microengineering—Descriptions, reviews, purchase source. [Amazon.com]
- Nano: The Emerging Science of Nanotechnology—Descriptions, reviews, purchase source. [Amazon.com]
- Nano: The Emerging Science of Nanotechnology—By Ed Regis; Little Brown & Co., 1996, ISBN 0316738522. Overview of short, rapid history of this topic, proponents and opponents, prospects. [Time Warner Bookmark]
- Nanocosm: Nanotechnology and the Big Changes Coming from the Inconceivably Small—By William Illsey Atkinson; AMACOM, 2003, ISBN 0814471811. Lay description of topic, many opinions expressed, not as accurate as it could be. [American Management Association]
- Nanocosm: Nanotechnology and the Big Changes Coming from the Inconceivably Small—Descriptions, reviews, purchase source. [Amazon.com]
- Nanoelectromechanics in Engineering and Biology—Descriptions, reviews, purchase source. [Amazon.com]
- Nanoelectromechanics in Engineering and Biology—By Michael Pycraft Hughes; CRC Press, 2002, ISBN 0849311837. Nanotechnology depends on abilities to manage nanoscale objects via interactions of electric fields, nanoparticles, ambient molecules. [CRC Press]
- Nano-Engineering in Science and Technology: An Introduction to the World of Nano-Design—Descriptions, reviews, purchase source. [Amazon.com]
- Nano-Engineering in Science and Technology: An Introduction to the World of Nano-Design—By Michael Rieth; World Scientific Publishing, 2003, ISBN 9812380736. Technical introduction, starts with quantum mechanical treatment of many-particle problem; focus: molecular dynamics method. [publisher website]
- Nanomedicine, Volume 1: Basic Capabilities—By Robert A. Freitas Jr.; Landes Bioscience, 1999, ISBN 157059645X, volume 1 of 3. About applying nanotechnology to medicine. [Online]
- Nanomedicine, Volume 1: Basic Capabilities—Descriptions, reviews, purchase source. [Amazon.com]

- Nanoscience: Friction and Rheology on the Nanometer Scale—Information, purchase source. [Amazon.com]
- Nanoscience: Friction and Rheology on the Nanometer Scale—Edited by E. Meyer, T. Gyalog, R.M. Overney, K. Dransfeld; World Scientific Publishing, 1999, ISBN 9812380620. Technical introduction to nanometer-scale tribology via friction force probe microscopy. [publisher website]
- Nanostructured Materials—Edited by Jackie Yi-Ru Ying; Academic Press, 2001, ISBN 0127444513. Course book and compact reference for academic and industrial readers. [Amazon.com]
- Nanosystems: Molecular Machinery, Manufacturing, and Computation—By K. Eric Drexler; John Wiley & Sons, 1992, ISBN 0471575186. Large, pioneering technical and mathematical text, may be author's greatest contribution to this field. [publisher website]
- Nanosystems: Molecular Machinery, Manufacturing, and Computation—Descriptions, reviews, purchase source. [Amazon.com]
- Nanotech Book Reviews—Steve Lenhert reviews his two favorite nanotech books, Travels to the Nanoworld and Nanotechnology.
- Nanotechnology—Edited by Gregory Timp; Springer-Verlag, 1999, ISBN 0387983341. Undergraduate level survey of machinery and science of nanometer scale, many images, 22 authors of many disciplines. [Springer-Verlag NY]
- Nanotechnology—Descriptions, reviews, purchase source. [Amazon.com]
- Nanotechnology: A Gentle Introduction to the Next Big Idea—By Mark Ratner, Daniel Ratner; Prentice Hall PTR, 2002, ISBN 0131014005. For non-technical readers, simple, brief, survey, little math; shows how topic works, what's new and next, why it may be next $1 trillion industry. [Prentice Hall PTR]
- Nanotechnology: A Gentle Introduction to the Next Big Idea—Descriptions, reviews, purchase source. [Amazon.com]
- Nanotechnology: Basic Science and Emerging Technologies—Descriptions, reviews, purchase source. [Amazon.com]
- Nanotechnology: Basic Science and Emerging Technologies—Edited by Mick Wilson, Kamali Kannangara, Geoff Smith, Michelle Simmons, Burkhard Raguse; CRC Press, 2002, ISBN 0849311837. Accessible to nonspecialists, engineers, scientists outside the field, even undergraduates. [CRC Press]
- Nanotechnology: Molecular Speculations on Global Abundance—Edited by B.C. Crandall; MIT Press, 1996, ISBN 0262531372. Accessible to wide

audience, with references to technical literature, shows wide range of uses of nanotechnology. [MIT Press]

- Nanotechnology: Molecular Speculations on Global Abundance—Descriptions, reviews, purchase source. [Amazon.com]
- Nanotechnology: Research and Perspectives—Edited by B.C. Crandall, James Lewis; MIT Press, 1992, ISBN 0262031957. Overview of presently understood paths to molecule scale engineering. [MIT Press]
- Nanotechnology: Research and Perspectives—Descriptions, reviews, purchase source. [Amazon.com]
- Nanotechnology Research Directions: IWGN Workshop Report: Vision for Nanotechnology in the Next Decade—Edited by R.S. Williams, P. Alivisatos, M.C. Roco; Kluwer Academic Publishers, 2000, ISBN 0792362209. Report based on consensus views of government, academia, private sector, by Interagency Working Group on Nano Science, Engineering, and Technology. [Amazon.com]
- Nanotechnology: Towards a Molecular Construction Kit—By Arthur ten Wolde; New World Ventures, 1998, ISBN 908044961X. Describes topic; where physical, chemical and biotic approaches meet; possible technologies for nano-fabrication, possible effects. [Amazon.com]
- The Next Big Thing Is Really Small: How Nanotechnology Will Change the Future of Your Business—Descriptions, reviews, purchase source. [Amazon.com]
- The Next Big Thing Is Really Small: How Nanotechnology Will Change the Future of Your Business—By Jack Uldrich, Deb Newberry; Crown Publishing Group, 2003, ISBN 1400046890. Introductory lay treatment, very useful for understanding day-to-day and business implications. [Random House]
- Our Molecular Future: How Nanotechnology, Robotics, Genetics and Artificial Intelligence Will Transform Our World—By Douglas Mulhall; Prometheus Books, 2002, ISBN 1573929921. Examines vast potential of new technologies to help us cope with many problems. Site has table of contents, excerpts, news release. [Prometheus Books Publishers]
- Our Molecular Future: How Nanotechnology, Robotics, Genetics and Artificial Intelligence Will Transform Our World—Descriptions, reviews, purchase source. [Amazon.com]
- Physical Properties of Carbon Nanotubes—Description, information, contents, sample chapters. [World Scientific Publishing]

- Physical Properties of Carbon Nanotubes—By Riichiro Saito, Gene Dresselhaus, Mildred S. Dresselhaus; Imperial College Press, 1998, ISBN 1860940935. Introductory textbook for graduate students, researchers in condensed matter and solid state physics. [publisher website]
- Physical Properties of Carbon Nanotubes—Descriptions, reviews, purchase source. [Amazon.com]
- Prospects in Nanotechnology: Toward Molecular Manufacturing—Edited by Markus Krummenacker, James Lewis; John Wiley & Sons, 1995, ISBN 0471309141. Result of Foresight Institute 3rd nanotechnology conference; introduction and overview, then topical presentations. [publisher website]
- Prospects in Nanotechnology: Toward Molecular Manufacturing—Descriptions, reviews, purchase source. [Amazon.com]
- Quantum Dots—Information, purchase source. [Amazon.com]
- Quantum Dots—Edited by Elena Borovitskaya, Michael S. Shur; World Scientific Publishing, 2002, ISBN 9810249187. Technical review of all current aspects of quantum dot systems, theory, and technology. [publisher website]
- Semiconductors for Micro- and Nanotechnology: An Introduction for Engineers—Descriptions, reviews, purchase source. [Amazon.com]
- Semiconductors for Micro- and Nanotechnology: An Introduction for Engineers—By Jan G. Korvink, Andreas Greiner; John Wiley & Sons, 2002, ISBN 3527302573. Textbook on properties and behavior of semiconductors, how to use them to design and make micro- and nanosystems. [publisher website]
- Travels to the Nanoworld: Miniature Machinery in Nature and Technology—Descriptions, reviews, purchase source. [Amazon.com]
- Unbounding the Future: The Nanotechnology Revolution—By K. Eric Drexler, Chris Peterson, Gayle Pergamit; William Morrow and Company, Inc., 1991; Quill (reprint), 1993, ISBN 0688125735. Clear, nontechnical introduction to molecular manufacturing: what it is, and may make possible, dangers and opportunities. [Foresight Institute, Online]
- Unbounding the Future: The Nanotechnology Revolution—By K. Eric Drexler, Chris Peterson, Gayle Pergamit; Quill (reprint), 1993, ISBN 0688125735. [Amazon.com]
- Understanding Nanotechnology—Edited by Scientific American; Warner Books, 2002, ISBN 0446679569. Accessible overview of this topic, and many profound applications. [Time Warner Bookmark]

- Understanding Nanotechnology—Descriptions, reviews, purchase source. [Amazon.com]
- Review of Nanocosm—By Chris Phoenix. This anti-Drexler book reports on nanotech but suffers from many technical inaccuracies and distortions of work of nanotech pioneers. [KurzweilAI.net] (June 6, 2003)

RECENT NEWS ON NANO TECHNOLOGY & NANO LABS

Nanowires may provide innards for quantum computers

SEMICONDUCTING nanowires could one day store and process data in quantum computers.

A team of researchers led by Silvano De Franceschi of the Delft University of Technology in the Netherlands made nanowires out of indium arsenide, each about 100 nanometres in diameter and 100 times as long. By attaching aluminium electrodes to the ends of the wires, they could probe the wires' behaviour at temperatures close to absolute zero. At such temperatures aluminium becomes superconducting, able to conduct electrical current with zero resistance. The team found that the superconducting electrodes induced similar behaviour in the nanowires because the superconducting properties leaked into the wire - the so called "proximity effect". By applying a voltage to the substrate on which the nanowires rested they could alter the strength of the proximity effect and so turn the superconductivity on and off at will (*Science*, vol 309, p 272). Because superconductivity is an inherently quantum effect, superconducting nanowires could form part of the quantum circuitry necessary for a future quantum computer. They could store information, or help read out the result of a calculation.

3D barcodes to identify stolen valuables

DIAMONDS and valuable works of art could be protected against theft using a microscopic barcode that stores encrypted information about the provenance of the items, making ownership easy to prove if they are stolen.

The barcode, which takes the form of a cube 30 micrometres across, is being developed by a team at the National Physical Laboratory (NPL) at Teddington, near London. The cube is made of silicon coated with a 100-nanometre-thick layer of polymethyl methacrylate, a transparent plastic. It can be attached to hard surfaces using adhesive, or woven into the canvas of paintings.

To create the barcode an electron-beam lithograph drills 90,000 small squares into the plastic coat of each face at five different depths. The position and depth of each square is unique, so data can be encrypted using a key-based code and stored digitally. The cube is scanned line by line using an electron force microscope, which can detect differences in the depth of the squares. This scanning process takes around a minute, says Alexandre Cuenat, one of the team of researchers that developed the cube.

The technology has huge potential for storing information securely, says Cuenat. "You could get two copies of the bible, the King James version, on the sharp end of a pin."

However, in practice storing information on the provenance of an object would not require anything like this capacity.

The nanobarcode has three levels of security to protect objects against theft and counterfeit. First, thieves and fences are unlikely to realise that the item is protected. "You cannot see or feel the cube, even if you roll it between your finger and thumb," says Cuenat. Second, most fraudsters are unlikely to have the specialist equipment to read or write the nanobarcode, and so will be unable to duplicate it. And finally the encryption will be practically unbreakable.

"Theoretically all codes can be broken," says David Mendels, who led the NPL team. You just have to test all the different permutations. In this case, it would be possible to make the encryption extremely complex because of the cube's high storage capacity. "You'd need a computer doing a million calculations a second and it would still take more than 187 billion years," says Mendels. For day-to-day use, the encryption could be far less complex and still be virtually unbreakable, says Cuenat.

Billions of tiny mechanical levers could be used to store songs on future MP3 players and pictures on digital cameras.

As bizarre as the idea might sound, researchers at a Dutch company have already demonstrated that miniscule mechanical switches can be used to store data using less power than existing technologies and with greater reliability.

Nanomech memory, developed by Cavendish Kinetics in the Netherlands, stores data using thousands of electro-mechanical switches that are toggled up or down to represent either a one or zero as a binary bit. Each switch is a few microns long and less than a micron wide - roughly one-hundredth the width of a human hair.

Existing computer memory typically stores data as an electrical or magnetic charge. Cavendish Kinetics claims Nanomech memory can read and write data using 100 times less power than such systems, and works up to 1000 times faster. It is also much more resilient to both temperature and radiation, the company claims.

Intermolecular forces

Nanomech memory incorporates hundreds of thousands of conductive metal levers, each just a few microns long. These are created by lithographically etching a design onto a template and then chemically dissolving away unwanted layers.

Introducing a miniscule voltage to an electrode below a lever causes it to bend forwards until it makes contact. Thanks to intermolecular forces on this scale, once flipped, a lever will also maintain its position, even when the voltage is switched off. The trick can be reversed by applying voltage to an electrode on the other side of the lever. And the state of each switch flips can be sensed easily by the electrodes themselves.

Repeating the feat across thousands of switches makes it possible to store something as complex as a computer program in Nanomech memory.

Cavendish Kinetic has so far developed a unit capable of storing 256 kilobytes of data, or around two million individual bits. This is suitable for simple microcontrollers - the tiny computers used to operate commonplace electric motors found, for example, in cars and consumer electronics products.

Gigabyte memory

But eventually the company hopes to stack many more levers together, boosting memory capacity to several gigabytes. "That's the long term aim," says Charles Smith at Cambridge University, UK, who is a chief technology officer at Cavenish Kinetics. "We want to put millions and millions on an individual chip."

This should make Nanomech memory a viable solution for portable music players and digital cameras. And it could improve battery life dramatically.

"The big growth area for large scale non-volatile memory devices is portable consumer applications, such as iPods and cameras," says Jim Miles, another nanotechnology researcher at Manchester University, UK. "And for these devices, power consumption is the big issue."

Miles, however, points out that it is vital to be able to repeatedly read and write to this type of memory. "It's important to know rewrite speeds as well as how many rewrite cycles they can survive," he says. Smith says the first increased capacity Nanomech chips should be available by the end of 2006. He adds that the memory technology's resilience to radiation should also make it suitable for use aboard satellites and other spacecraft. This is because cosmic radiation can cause space-borne computers to suddenly malfunction by flipping bits stored in memory.

World's smallest toothbrush scrubs capillaries clean

Nano-sized toothbrushes that can clean very small surfaces have been developed by researchers.

Fabricated out of millions of carbon nanotubes, the minuscule brushes could even paint the inside of capillaries thinner than a human hair.

But besides being small, the nanotubes have other advantages over traditional bristles, says one of their creators Pulickel Ajayan at Rensselaer Polytechnic Institute in Troy, New York. The materials typically used for making brush bristles include animal hairs, synthetic polymer fibres and metal wires. But each has its limitations. Metals corrode and weaken, hair is not very strong and synthetic fibres melt.

Carbon nanotubes, on the other hand, are 30 times stronger than steel, yet five times less dense. They are highly elastic, resistant to heat, have large surface areas and even conduct electricity. The latter property makes them highly suitable for the contact brushes used in electric motors, says Ajayan.

He and Anyuan Cao at Rensselaer, in collaboration with colleagues at the University of Hawaii in Honolulu, US, have designed brushes consisting of a silicon carbide fibre base on which the carbon nanotubes are grown in a single row of bristles like a toothbrush, or in groups of bristles more like a toilet brush. One end of the base is coated with gold, which acts like a "handle" and also inhibits nanotube growth at that end. The dimensions can be varied dramatically, says Ajayan, but typically the bundle of nanotubes is no smaller than a few micrometres in diameter.

The brushes could have a variety of applications. They have already been used to sweep particles measuring just 50 nanometres in diameter off a plain surface, and off another surface pitted with microscopic grooves. The researchers have also attached the gold handle of a three-pronged brush to the shaft of a small electric motor. This allowed them to rotate the bristles inside a 300-micrometre-wide capillary tube, cleaning contamination off its walls. Then, by coating the brush with a red dye, they were able to paint the inside of the cavity using the same approach.

The brushes' ability to clean small particles appears to be extremely good, says Ajayan, although he is still not sure why. "Perhaps the cleaning occurs by electrostatic interactions, and since the surface area of the nanostructures is large, the cleaning is possibly more effective." The aim now is to apply the brushes to more specific microelectronic and biomedical applications.

Journal reference: *Nature Materials* (DOI: 10.1038/nmat1415)

Harmful bacteria shown up by nanoparticles

A new nanoparticle test for dangerous bacteria such as *Escherichia coli* O157:H7 is so sensitive it can detect a single bacterial cell within minutes. The food industry, medicine and the fight against bioterrorism could all eventually benefit from it, researchers say.

Even a few cells of the *E. coli* strain in food can be dangerous so it is important to be able to detect them in low numbers. Current tests, however, need a higher number of bacteria to be present before they can detect it, which can lead to long delays.

The new test, developed by Weihong Tan and colleagues from the University of Florida, US, could offer a sharper and faster way of detecting contamination. "If you can give us one bacterial cell in a sample then we can detect it," says Lisa Hilliard, one of the team. "Most people have to grow it and then detect it."

The whole test can be carried out in just 20 minutes, compared with up to 48 hours for conventional tests. "Other tests could be as sensitive but you would need to perform an enrichment first," says Mike Peck from the Institute of Food Research in the UK. "You would have to pop the food into a growth media for 24 hours," he explains.

Shelved beef

Waiting a day or two for the results can be costly and inconvenient, says Andrew Brabban, who works on *E. coli* at the Evergreen State College, Washington. "One of the basic problems at the moment is what is called 'Test and Hold'. Samples are held at US plants until they are shown to be free of O157:H7. This is obviously expensive for the industry, having large quantities of beef as shelved stock," he told New Scientist.

The new test consists of silica nanoparticles, each housing thousands of fluorescent dye molecules, and each attached to an antibody for a given bacterium.

The nanoparticles are added to a solution of the test sample, such as ground beef. If the bacterium sought is present the nanoparticles will quickly attach to it. The sample is then separated by weight in a centrifuge. The target bacteria, being heavier than the nanoparticles, will separate away from them. But those dye molecules already attached will fluoresce in this heavier sample, identifying the bacteria.

Diluted samples

The new test differs from other dye tests in that thousands of dye molecules will fluoresce if only one bacterium is present because they are all attached to the same antibody. Other tests contain only a few dye molecules for each test antibody used so that one bacterial cell will not cause enough fluorescence to be observed. Using samples so diluted that only one in four of them contained any bacteria at all, the researchers showed their test could detect a single bacterial cell.

Although the test was developed using *E. coli* O157:H7, it could be adapted to many different bacteria. The group are already looking at ways to identify more than one type of bacteria at a time, using different coloured dyes for each.

If this can be achieved quickly and accurately it could have great potential. "The need to detect single cells is real," says Brabban. "Any method that is fast, accurate and has [such a sensitive] detection level would certainly be very useful," he says.

Journal reference: *Proceedings of the National Academy of Sciences* (DOI: 10.1073/pnas.0404806101)

Buckyballs made safer for humans

Carbon molecules called "buckyballs" - which hold great promise for nanotechnology - but have been shown to harm fish have been made safer by scientists.

The soccer-ball-shaped carbon nanoparticles were shown to cause brain damage in fish and kill water fleas in a study in March 2004. But now a team at Rice University in Houston, Texas, US, has come close to understanding how buckyballs – more formally known as fullerenes - kill cells and how their toxicity can be lowered in human cells.

Although the toxic nature of the carbon-60 nanoparticles may be useful in medicine, for example in fighting cancer, there are concerns that their potentially widespread use in fuel cells, drug delivery and cosmetics could mean they find their way into the environment, and so into animals and humans.

"There are a couple of different manufacturers that will, and are, mass producing fullerenes," says Christie Sayes, one of the team. "They could make it into consumer based products: fuel cells and batteries or make-up," she says.

Ruptured cells

The team, led by Vicki Colvin, looked at the effects of buckyballs on human cells. They found that even at quite low concentrations in water the buckyballs killed human skin cells. However, when they attached chemical groups such as hydroxyl groups to the buckyballs, their toxicity was greatly reduced.

The higher the number of groups that were attached, the less toxic the fullerenes became, so that a buckyball with 24 hydroxyl groups attached showed a toxicity seven orders of magnitude lower than the original buckyball.

The researchers believe the buckyball is toxic because in water it leads to the formation of an oxygen free radical which reacts with lipid molecules forming the cell membrane surrounding a cell. This causes the lipid molecules in turn to become free radicals – and when these try to interact with the water outside the cell, the membrane ruptures leading to cell death.

Health threat

The team were able to show the hole in the membrane using dyes that only fluoresce when in the cell. Dyes that are usually too big to enter a cell fluoresced after the buckyballs had been added, showing the permeability of the cell had increased to such an extent that even large molecules could move freely through the membrane.

More research is needed to determine whether the same result would be seen in the body, says Sayes. "Cells in culture act differently to those in the body," she says.

The research could eventually be use to make buckyballs safer, so that they do not cause an environmental or health threat as they become more commonly used. Or even to make them more toxic for certain medical treatments.

Eva Oberdõrster at Southern Methodist University in Dallas, Texas, who led the buckyball study in fish, called the new work "an excellent first step" in reducing the toxicity of buckyballs.

Journal reference: *Nano Letters* (in press)

U-turn on goo

GREY goo is no more. Eric Drexler, the futurist who dreamed up the vision of self-replicating nanomachines spreading across the planet has publicly repudiated his idea.

In his 1986 book Engines of Creation, Drexler suggests that microscopic machines might one day be able to manipulate individual molecules to build any desired structure. If such a machine contained its own blueprint and could scavenge raw materials, it could take over the planet in a chain reaction of self-replication, he warned.

"The self-replicating machine idea has had a profound effect on nanotechnology, or rather its perception - mostly for the bad," says Mark Welland, editor-in-chief of the journal Nanotechnology.

Now, in an article in the June issue of Nanotechnology, Drexler says that nanofactories will be desktop-sized machines that can manipulate large numbers of molecules. Although such a factory could be directed to build a copy of itself, says Drexler, it would not be able to do so on its own. And neither would it be able to spread around the planet, Hollywood style, eating everything its path.

Carbon nanotubes used in computer and TV screens

The next generation of computer and television screens could be built using carbon nanotubes. Next week a prototype high-definition 10-centimetre flat screen made using this technology will be launched in Boston at the Society for Information Display conference.

The new screen, called a nano-emissive display or NED, is made from two sheets of glass, one covered by a layer of nanotubes standing on end, the other by a layer of blue, red or green phosphors similar to those used in conventional cathode ray tube screens. When charged, the nanotubes direct electrons at the phosphors, making them light up. Because the electrons have only a short distance to travel, even a 105-centimetre NED would use relatively little power, says maker Motorola. A screen that size will also have a wide viewing angle and could sell for less than $400, the company claims.

Building nanomachines out of living bacteria

LIVE bacteria could one day act as reconfigurable components for nanoscale electronic circuits, or even a scaffold for building nanomachines.

"Nature has developed these fantastic building blocks," says Robert Hamers of the University of Wisconsin-Madison. "Our approach is to simply grab onto them very gently." His team have been using electrodes to manipulate individual bacterial cells, they will report in a future issue of Nano Letters.

Their chosen bacterium was Bacillus mycoide, 5 micrometres long and 0.8 wide - large enough to be visible under an optical microscope. The trick is to apply an electric field to the microbes, so they become polarised and stick to the electrodes. But the idea should also work with smaller bacteria, as a change in the electrical current of the electrode would show they were attached.

The researchers even persuaded the bacteria to form a conducting bridge between two electrodes, a knack that could perhaps lead to reconfigurable nano-circuitry. "The idea of using cells as part of a more complex electronic circuit is very important," says Hamers.

In his experiments the bacteria attached themselves to a long electrode before being shunted along its length by the motion of the solution in which they are suspended. When the bacteria reach a junction with another electrode, the polarities of the two electrodes securely fix the bugs in place (see Graphic).

At the moment, nanostructures have to be put together manually. But it might be possible to automate the process using bacteria, as components tagged with particular biological molecules will stick to complementary surface proteins on the bacteria. Another use for Hamers-type electrodes would be in biosensors that could detect biological agents such as anthrax from changes in an electrode's current as spores become attached.

Mechanical chip promises huge data storage

A super-dense memory chip that stores data in the form of nanoscale holes in a plastic film has made its public debut at the CeBIT electronics exhibition in Hanover, Germany.

Storing data in the form of holes is not new - CDs use pits in a polycarbonate disc, for instance, and 19th-century looms stored patterns on punched cards. But the "Millipede" technology from IBM's Zurich lab promises very high capacity thanks to its use of holes just 10 nanometres wide. This means that a square chip measuring 2.4 centimetres on a side should be able to store 125 gigabytes, says the company, equivalent to 25 DVDs.

The Millipede chip achieves this by having an array of tens of thousands of silicon cantilevers, explains IBM researcher Evangelos Eleftheriou. Each has a pointed tip that writes data by poking holes - representing a digital 0 or 1 - in the soft polymer below. The cantilever also reads the data when needed, by measuring a change in its electrical resistance when it drops into a hole.

Electromagnetic actuators within the chip package move the polymer film beneath the cantilevers so that each tip can read and write within a 100-micrometre-square area. Data is erased using a heater in each cantilever which melts the polymer locally, smoothing the pits over for re-use.

Partner search

That is the theory, but to make Millipede a commercial reality IBM admits it needs an industrial partner with expertise in the manufacture of microelectromechanical systems (MEMS).

"We have no MEMS product line at IBM. Companies who make accelerometers or microscale actuators will have the kind of production capabilities Millipede needs," says Eleftheriou. The technology's appearance at CeBIT was geared towards finding that partner.

At the show, IBM used a video microscope to demonstrate the micro-cantilevers going about their work on a 10 gigabyte version of a Millipede chip.

If IBM can commercialise the memory chip and get it reading and writing data at speeds of 20 to 30 megabits per second - like today's "flash" memory chips - it hopes the technology could form the heart of future digital cameras, cellphones and USB memory sticks.

However, some analysts are sceptical of Millipede's prospects, questioning the need for a mechanical memory device when consumer electronics firms already have electronic, optical and magnetic technologies in use.

Nanoscopic 'ruler' could provide microchip benchmark

A nanoscopic measuring device that uses atomic lattices to gauge tiny distances could enable microelectronics engineers to build better components, say US scientists.

The "nanoruler" was developed by researchers at the US government's National Institute of Standards and Technology (NIST) in Maryland, and industry collaborators. They say it will enable researchers and engineers working on nanoscale objects to calibrate their instruments more precisely, resulting in greater accuracy in their work.

To construct the instrument, NIST researcher Michael Cresswell says the team exploited the way in which silicon atoms align themselves into uniform lattice structures.

They had to etch the raw silicon in the correct orientation to ensure the lattices were perpendicular to the top and bottom surfaces. If not, the distance between each lattice would not be standard.

The nanoscopic slab can provide precise measurements by counting the number of silicon lattice lines crossed by an object. An advantage silicon is that the width of each atomic lattice is known to within about one-tenth of a nanometre, says Cresswell, at 0.3 nanometres.

Nanorulers varying in size from 40 to 275 nanometres can be constructed. These could, for example, be used by nano-engineers to improve the calibration of instruments such as atomic force microscopes, scanning tunnelling microscopes and electron microscopes.

Especially useful

Researchers currently use a variety of materials to calibrate their instruments, the best of which is still seven times less accurate than the silicon ruler, Cresswell told *New Scientist*. He adds that the nanoruler could prove especially useful in microelectronics fabrication, where engineers check the dimensions of components after each step of their work.

Peter Hatto, chairman of the European Committee for Standardization's working group on nanotechnology, says a precise standard for nano-measuring would be a very useful research tool.

"You really need to use a fundamental measurement standard but you can't use the metre without an artefact to calibrate against," he says. The NIST nanoruler has the advantage that the properties of silicon are already well understood, he adds.But Hatto also notes that it will be crucial to provide researchers with easy access to the material to ensure it is used.

The ruler was developed by NIST in collaboration with three US companies; Sematech, in Texas, US, and VLSI Standards and Accurel, both based in California, US.

Mechanical chip promises huge data storage

A super-dense memory chip that stores data in the form of nanoscale holes in a plastic film has made its public debut at the CeBIT electronics exhibition in Hanover, Germany.

Storing data in the form of holes is not new - CDs use pits in a polycarbonate disc, for instance, and 19th-century looms stored patterns on punched cards. But the "Millipede" technology from IBM's Zurich lab promises very high capacity thanks to its use of holes just 10 nanometres wide. This means that a square chip measuring 2.4 centimetres on a side should be able to store 125 gigabytes, says the company, equivalent to 25 DVDs.

The Millipede chip achieves this by having an array of tens of thousands of silicon cantilevers, explains IBM researcher Evangelos Eleftheriou. Each has a pointed tip that writes data by poking holes - representing a digital 0 or 1 - in the soft polymer below. The cantilever also reads the data when needed, by measuring a change in its electrical resistance when it drops into a hole.

Electromagnetic actuators within the chip package move the polymer film beneath the cantilevers so that each tip can read and write within a 100-micrometre-square area. Data is erased using a heater in each cantilever which melts the polymer locally, smoothing the pits over for re-use.

Partner search

That is the theory, but to make Millipede a commercial reality IBM admits it needs an industrial partner with expertise in the manufacture of microelectromechanical systems (MEMS).

"We have no MEMS product line at IBM. Companies who make accelerometers or microscale actuators will have the kind of production capabilities Millipede needs," says Eleftheriou. The technology's appearance at CeBIT was geared towards finding that partner.

At the show, IBM used a video microscope to demonstrate the micro-cantilevers going about their work on a 10 gigabyte version of a Millipede chip.

If IBM can commercialise the memory chip and get it reading and writing data at speeds of 20 to 30 megabits per second - like today's "flash" memory chips - it hopes the technology could form the heart of future digital cameras, cellphones and USB memory sticks.

However, some analysts are sceptical of Millipede's prospects, questioning the need for a mechanical memory device when consumer electronics firms already have electronic, optical and magnetic technologies in use.

The CeBIT exhibition runs from 10 to 16 March.

Charge a battery in just six minutes

A rechargeable battery that can be fully charged in just 6 minutes, lasts 10 times as long as today's rechargeables and can provide bursts of electricity up to three times more powerful is showing promise in a Nevada lab.

New types of battery are badly needed. Nokia's chief technologist Yrjö Neuvo warned last year that batteries are failing to keep up with the demands of the increasingly energy-draining features being crammed into mobile devices (New Scientist print edition, 28 February 2004).

The highest energy-per-weight ratio in today's batteries is provided by lithium-ion (Li-ion) batteries. They are also cheaper in terms of energy delivered per unit of weight than alternative types of battery such as nickel-metal-hydride (NiMH) and nickel-cadmium (Ni-Cd) types. But Li-ion cells have their drawbacks too. They eventually wear out, and they cannot discharge energy quickly enough for applications requiring power surges, such as camera flashguns and power tools.

They may soon be able to. Altair Technologies of Reno has created a new type of Li-ion cell in which the anode has an exceptionally high surface area. This allows electrons to enter and leave it quickly - making fast recharging possible and providing high currents when needed.

Chemical tricks

Li-ion batteries work by forcing lithium ions from a lithium cobalt oxide cathode to migrate to a carbon anode via an electrolyte solution. Altair's patented modification is to make the anode surface out of lithium titanate nanocrystals, using chemical tricks to give it a surface area of about 100 square metres per gram, compared with 3 square metres per gram for carbon.

The firm is keeping the chemistry that allows it to do this pretty close to its chest for commercial reasons. But the patent (US 6689716) reveals that the increased surface area is achieved using a carefully controlled sequence of evaporative steps when making the lithium titanate crystals.

The high current that this modified electrode is able to carry means power-hungry devices can be installed in mobile phones, which until now have been denied them. For instance, camera phones might now have enough power to run a flashgun.

Longer lifespan

Altair says the battery will have other advantages, too. The crystalline surface of a carbon anode is susceptible to damage by the repeated temperature changes that occur as the battery is used and recharged. This limits its life to around 400 charging cycles.

The more rugged lithium titanate anode should make it possible to recharge the battery as many as 20,000 times says Roy Graham, development director at Altair. A longer lifespan should also be better for the environment, he says. "The continual use of polluting cobalt oxides is questionable."

Altair plans to develop its batteries for power tools, which have till now required more expensive Ni-Cd or NiMH batteries to provide the large currents these devices need. The company hopes to license its technology to major battery-makers, who could have the device on the market in two years' time. Altair says it eventually wants to produce batteries for a broad range of devices, from phones to hybrid electric vehicles.

Nanotechnology may create new organs

Scientists have built a minute, functioning vascular system - the branching network of blood vessels which supply nutrients and oxygen to tissues - in a significant step towards building whole organs.

Conventional tissue engineering methods have successfully grown structural tissues such as skin and cartilage in the lab. But not being able to create the supporting vascular system has proved a major stumbling block preventing scientists from creating large functioning organs such as liver or kidneys.

Now, researchers from Massachusetts Institute of Technology and Harvard Medical School have used computers to design branching networks of venous and arterial capillaries, which start at three millimetres wide and reach a fineness of just 10 microns.

"We used living vessels as a guide to model factors such as the angle and size ratio between branching vessels. But we optimised our design to improve it," said lead researcher Mohammad Kaazempur-Mofrad, from MIT[1]s department of mechanical engineering and division of biological engineering.

Pig or rabbit

The networks were etched on to 15 centimetre-wide silicon wafers and the paths were then used as a mould to set a layer of biodegradable polymer. Two of these were then sealed together with a microporous membrane sandwiched between them, producing a mini artificial vascular system.

Endothelial cells - which are flat cells lining the walls of blood vessels in a single layer - were injected into the network on one side of the membrane and either liver or kidney cells were injected on the other side. The endothelial cells coated the inside of the polymer nanotubes. These nanotubes biodegraded to leave a living shell of vessels similar to a natural vascular network. This method would provide an efficient means of supplying the liver or kidney cells with enough oxygen and nutrients to survive.

The one-layer systems of kidney and liver cells were successfully implanted into rats for two weeks - 95 per cent of the cells survived.

"The next step is to work on bigger animals, such as a pig or rabbit, using more layers," Kaazempur-Mofrad told New Scientist. "Eventually, we want to be able to replace whole organs with several layers of these constructs. The critical mass for liver is one-third, probably 30 to 50 stacked layers."

"So in the next 10-15 years, we will hopefully have reached a point where we can do this procedure clinically in human patients," Kaazempur-Mofrad added.

The research was presented at the American Society for Microbiology Conference on Bio-, Micro-, and Nanosystems, in New York City on Tuesday.

Nanoparticles clearly finger the culprit

Oil-seeking nanoparticles could give police the clearest fingerprints yet, suggests new research. Law enforcement officers currently search for prints by dusting a crime scene with fluorescent powder. This sticks to the oily residue left by the fingertip, showing up the whorls and ridges. But sometimes the prints are not clear enough to finger a suspect. The new dust made of sticky nanoparticles could help. The powders used today work because oily prints have a natural tackiness. But the nanoparticle dust being developed at the University of Sunderland in the UK will actively seek out any oil. The nanoparticles are tiny glass spheres between 200 and 600 nanometres in diameter. As well as being speckled with a fluorescent dye, they are coated with hydrophobic molecules, which are repelled by water and attracted to oil. So they fix tightly to the fingerprint. Fred Rowell, who is leading the research, says the nanoparticles should pick out even the faintest of fingerprints because they stick to tiny traces of oil. And the prints should be much sharper, providing fine detail that can be crucial to identifying a print, such as how ridges in the print branch and finish.

Nano-transistor self-assembles using biology

A functional electronic nano-device has been manufactured using biological self-assembly for the first time.

Israeli scientists harnessed the construction capabilities of DNA and the electronic properties of carbon nanotubes to create the self-assembling nano-transistor. The work has been greeted as "outstanding" and "spectacular" by nanotechnology experts.

The push to shrink electronic circuits to ever smaller dimensions is relentless. Carbon nanotubes, which have remarkable electronic properties and only about one nanometre in diameter, have been touted as a highly promising material to help drive miniaturisation. But manufacturing nano-scale transistors has proved both time-consuming and labour-intensive.

The team, at the Technion-Israel Institute of Technology, overcame these problems with a two step process. First they used proteins to allow carbon nanotubes to bind to specific sites on strands of DNA. They then turned the remainder of the DNA molecule into a conducting wire.

Proof of principle

"DNA is very good at building things in molecular biology, but unfortunately, it does not conduct electricity. We had to get a metal conductor on the DNA," explains physicist Erez Braun, who led the research.

"This is spectacular work," says Cees Dekker, a nanoscience expert at Delft University in the Netherlands. "It demonstrates that it's possible to use biology to build an inorganic device that works."

"But while it is a first step towards molecular computing based on this type of DNA configuration, we are still many years way from large scale self-assembly electronic devices, such as computers," Dekker cautions.

Bacterial protein

Braun's team began their manufacturing process by coating a central part of a long DNA molecule with proteins from an *E. coli* bacterium. Next, graphite nanotubes coated with antibodies were added, which bound onto the protein.

After this, a solution of silver ions was added. The ions chemically attach to the phosphate backbone of the DNA, but only where no protein has attached. Aldehyde then reduces the ions to silver metal, forming the foundation of a conducting wire.

To complete the device, gold was added. This nucleates on the silver and creates a fully conducting wire. The end result is a carbon nanotube device connected a both ends by a gold and silver wire.

The device operates as a transistor when a voltage applied across the substrate is varied. This causes the nanotubes to either bridge the gap between the wires - completing the circuit - or not. Out of 45 nanoscale devices created in three batches, almost a third emerged as self-assembled transistors. They work at room temperature and the only restriction for future devices is that the components must be compatible with the biological reactions and the metal-plating process. The team have already connected two of the devices together, using the biological technique. "The same process could allow us to create elaborate self-assembling DNA sculptures and circuitry," says Braun. Journal reference: *Science* (vol 302, p 1380)

'Bio-barcoding' promises early Alzheimer's diagnosis

Combining magnetic and gold nanoparticles with strands of DNA could allow the early detection of Alzheimer's disease. If successful, future treatments could then be used to prevent symptoms from ever appearing.

Identifying and tracking Alzheimer's disease currently relies on brain imaging and psychological testing. A firm diagnosis can only be made by autopsy.

But recent studies have revealed several biochemical markers that may provide the basis for a living diagnosis. These include tiny proteins called amyloid-beta-derived diffusible ligands (ADDLs), which exist at elevated levels in the brains of Alzheimer's patients.

Dimitra Georganopoulou and her colleagues at Northwestern University in Chicago, Illinois, US, suspected ADDLs might also be present in the cerebrospinal fluid, which can be collected using a spinal tap. But the concentration of the proteins would be incredibly tiny, if they were there at all.

So the team turned to a recently developed technique called bio-barcoding, shown to be thousands of times more sensitive for protein detection than conventional tests. "It's the only tool in town with the necessary sensitivity," says team member Chad Mirkin, whose lab developed the technique.

Hugely amplified

The bio-barcoding technique involves the ADDLs becoming sandwiched between two different antibodies. One antibody is attached to a 1000-nanometre-diameter magnetic particle and the other is fixed to a 30-nanometer-diameter gold particle. The gold nanoparticle also has thousands of short segments of DNA attached to it - the so-called barcoding DNA.

Once the ADDLs are firmly attached to both types of the nanoparticle, they are drawn off using the magnetic particle. The separated mixture will contain all of the magnetic nanoparticles initially added but only those gold particles that became attached to the other side of an ADDL protein. The DNA snippets attached to the gold nanoparticles are then chemically released from the gold particles and counted, giving a hugely amplified signal of the original protein.

The Northwestern University team used the barcoding method on cerebrospinal fluid from 15 people with Alzheimer's disease - as confirmed after

Limit of sensitivity

The team will now test the technique on blood samples, which are far less difficult to collect, although the concentration of ADDLs is expected to be even lower in blood than in cerebrospinal fluid.

But the researchers are confident that the barcoding approach should still work. "We're nowhere near the limits of the test's sensitivity," says team member William Klein.

ADDLs have been shown to attack the connections between neurons, called synapses, far earlier than currently observable symptoms appear in patients.

So early detection could allow future treatments to prevent symptoms appearing in the first place, says Klein.

Experiments using antibodies against ADDLs have even reversed memory loss in transgenic mice, says Klein, a scientific advisor to Chicago-based Acumen Pharmaceuticals. The company is working in collaboration with US company Merck & Co., Inc. to develop Alzheimer's disease vaccines based on ADDL research.

Journal reference: *Proceedings of the National Academies of Sciences* (DOI: 10.1073/pnas.0409336102)

'Smart bombs' to deliver fatal blast to tumours

Nanoscale polymer capsules could one day be used to deliver chemotherapy direct to tumours, leaving adjacent tissue unscathed. The capsules would be designed to rupture when heated by a low-energy laser pulse, delivering their payload right where it is needed.

Anti-cancer drugs would be more effective, and the side effects less severe, if they could home in on a tumour and be delivered in a single burst. This would allow the drug to reach the concentrations needed to kill cancer cells, while minimising damage to surrounding tissue.

So Frank Caruso and his team at the University of Melbourne, Australia, are developing an ingenious way of doing this. Their trick is to enclose the drug in polymer capsules that are peppered with gold nanoparticles and attached to tumour-seeking antibodies.

When injected into the bloodstream, the capsules will concentrate inside tumours. When enough capsules have gathered there, a pulse from a near-infrared laser will melt the gold, which strongly absorbs near-infrared wavelengths. This will rupture the plastic capsules and release their contents.

Near-infrared laser

To make the capsules, the researchers repeatedly add the polymer to a suspension of drug particles around 1 micrometre wide until the polymer forms multi-layered spheres containing the drug payload. They then add gold particles around 6 nanometres in diameter to the mix, which become embedded in the polymer. Finally they add a lipid, which forms an outer layer, and the antibodies that will target tumour cells (see graphic).

In lab tests, the capsules were ruptured by a 10-nanosecond pulse from a near-infrared laser. While bulk gold has a melting point of 1064°C, gold nanoparticles melt at far lower temperatures - between 600°C and 800°C. The brief pulse was enough to melt the nanoparticles, which could be seen under an electron microscope to swell to up to 50 nanometres in diameter as they coalesced.

The pulse was too short to damage the contents of the capsule. Caruso showed that a lysozyme enzyme did not lose its activity after being released from the capsules in this way.

Safety limits

In clinical use, the laser would be able to penetrate a few millimetres of tissue. It could be shone through the skin, or be beamed inside the body via an endoscope. The 100 millijoules per square centimetre of infrared energy that would be needed to rupture the capsules is well within safety limits. "It's way below that used to remove tattoos," Caruso says.

Clinical use is still some way off, however, and even animal tests are several years away. The next refinement will be to make the capsules a lot smaller. Caruso plans to shrink them from around 1 micrometre in diameter to a couple of hundred nanometres by starting with smaller drug particles.

Caruso thinks his team's key innovation has been in making the capsules react to a laser that is harmless to the body. Gold usually absorbs light from the visible to ultraviolet part of the electromagnetic spectrum, which can burn tissue. But electromagnetic interactions between the gold nanoparticles in the capsules change the properties of the metal, making it absorb light from the near infrared instead. This light is the most transparent to tissue.

Observers are impressed. "This work is tremendously cool," says Matt Trau of the Nanotechnology and Biomaterials Centre at the University of Queensland in Brisbane. Photo-activating the capsules without damaging surrounding tissue is the particularly innovative part, he says.

Journal reference: *Advanced Materials* (vol 16, p 2184)

DNA 'Velcro' binds nanoparticles

Strands of DNA have been used to fasten - and then separate - nanoparticles in experiments that could lead to the development of fundamentally "self-constructing" materials.

Christof Niemeyer and colleagues at the University of Dortmund, Germany, used sections of artificially synthesised DNA to attach gold nanoparticles together before separating them again.

DNA consists of two complementary strands that bind together depending on the chemical "bases" on either strand. Attaching nanoparticles by appending them to complementary strands of DNA is a tried and tested technique, already used in some protein sensing systems. But, until now, no one has shown that it is possible to separate the strands again afterwards.

Selectively binding nanoparticles could provide a way to construct complex nanostructures piece by piece, using different DNA strands to add different nanoparticles. These materials could have novel electrical and optical properties that cannot currently be obtained using conventional chemistry.

Third strand

Being able to separate these materials again would offer even greater flexibility. Niemeyer says it could conceivably be used to modify nanostructures after construction. "It could be used to build [self-constructing] materials," he told New Scientist. "Although that is very far off at the moment."

The team uses artificial DNA, fabricated to have particular bases. Each gold particle - measuring around 15 nanometres (15 billionths of a metre) across - is attached using sulphur to the centre of a DNA strand. The strand extends out on either side.

Two gold particles are then joined together by adding a third strand of DNA that complements half of each gold-bound DNA strand and attaches to both, forming a bridge. This just left the problem of how to separate the gold particles once the bridge was firmly in place.

Peeling away

So Niemeyer's team developed the bridging strand of DNA to have one end longer than the other with the tip of the long end refusing to attach to the gold-bound strands, leaving it loose.

The trick, finally, was to use an additional fourth piece of DNA to bind to this loose end of the bridging strand before "peeling" it away completely - like pulling off a piece of Velcro - leaving the gold particles separated once again.

Chad Mirkin, who pioneered the use of DNA as a nanoscale construction material says the approach is interesting, although the potential applications remain unclear.

"It's a step towards creating a structure where you could have triggerable changes," says Mirkin, now based at Northwestern University in Chicago, US. "But it's not yet clear where these would be useful."

Journal reference: *Angewandte Chemie* (vol 43, p 6469)

The Global Market for High-Brightness LEDs

This new report, brought to you jointly by NanoMarkets and CIR, analyzes the world wide market for high brightness LEDs (HB-LEDs). Although there have been recent rumors of industry maturity in this sector recently, our new study finds that HB-LEDs continue to grow rapidly, both in terms of revenues and in terms of new applications. This report is geared to the needs of both managers and investors who must understand the future prospects for the HB-LED markets and where the emerging business opportunities are to be found. Among the areas on which this report provides in-depth guidance are the following:

.Key design and manufacturing improvements that are enabling HB-LED manufacturers to distinguish themselves in the market and improve yields

.The impact of new materials platforms such as quantum dots and silicon photonics for the HB-LED market

.Emerging applications in signage, intelligent transport systems, rural lighting and dentistry and how HB-LEDs are enabling entirely new kinds of products to be designed and manufactured

.The changing geography of HB-LED markets. In which part of the world are HB-LEDs being bought? In which parts of the world are they being manufactured?

.Strategies and structure. In this report, we critically appraise the strategies of the major HB-LED manufacturers including their manufacturing and outsourcing strategies and ask the questions — is there room for start-ups in HB-LED space anymore? And, will there be consolidation in the HB-LED busines?

CIR has been tracking the optoelectronics business for more than two decades while NanoMarkets has produced over a dozen reports that offer a unique perspective on the impact of new materials on electronics and semiconductor markets. This report was based on interviews conducted by an international team carrying out interviews on three continents and will be invaluable reading for manufacturers of electronics and lighting products of all kinds as well as the HB-LED manufacturers themselves — and, of course, investors. The report includes detailed forecasts broken out by type of product and application.

HP Quantum Science Researchers Win Two More Awards

Researchers in molecular electronics at HP Labs, already the recipients of more than a half-dozen awards, have been honored by The Wall Street Journal and by Small Times magazine.

The Wall Street Journal gave a runner-up Global Technology Innovation Award in its semiconductor category to the team's molecular crossbar latch, a technology that may be able to replace transistors – the fundamental building blocks of computers for the past 40 years – leading to a new way to build computers in the future.

In addition, Phil Kuekes, senior computer architect, was named Best of Small Tech researcher of the year award by the nanotechnology publication Small Times for his contributions to the crossbar latch and other advances.

Kuekes accepted the award at the publication's NanoCommerce conference in Chicago Nov. 2.

The Wall Street Journal solicited nominations from all over the world through advertisements in the paper and through the Innovation Awards Web site. The newspaper received some 750 applications from approximately two dozen countries. Applicants completed a rigorous form that requested extensive details relating to their entry and why it constituted a true innovation.

Wall Street Journal editors then screened the applications and selected 104 semifinalists. Journal editors looked for innovativeness, clarity of explanation and whether the innovation was covered by patents or had achieved some kind of track record.

"Innovation is the key for any company to have continued success, and through these awards The Wall Street Journal recognizes these technological breakthroughs throughout the world," said Karen Elliott House, Journal publisher and senior vice president, Dow Jones & Company. "This year's winners showed the creative thinking that has allowed them to rise to the top of their respective industries."

Wake Forest Improve Efficiency of Organic or Flexible Solar Cells

At a time when oil prices are reaching record highs and people are bracing for winter heating bills, researchers at *Wake Forest University's Center for Nanotechnology and Molecular Materials* have made significant strides in improving the efficiency of organic or flexible solar cells.

Traditional silicon solar panels are heavy and bulky and convert about 20 percent of the light that hits them to useful electrical power. For years, researchers have worked to create flexible, or "conformal," organic solar cells that can be wrapped around surfaces, rolled up or even painted onto structures, but the best scientists have been able to do is about 3 percent efficiency, until now.

Researchers at Wake Forest, with the help of researchers at New Mexico State University, have achieved an efficiency rate for organic solar cells of almost 6 percent. In order to be considered a viable technology, the solar cells must be able to convert about 10 percent of the energy in sunlight to electricity. Wake Forest researchers hope to reach 10 percent by October 2006, said David Carroll, director of the nanotechnology center at Wake Forest.

"The consumer market would be really open to having these conformal systems if you could, for instance, roll them up and put them away," said Carroll, who is also an associate professor in Wake Forest's physics department. "Imagine a group of hikers with a tent that when you unrolled the tent and put it up, it could generate its own power. Imagine if the paint on your car that is getting hot in the sun was instead converting part of that heat to recharge your battery."

Carroll said flexible, organic solar cells also offer several possibilities for military use.

"The military would obviously want something like that because you could only put maybe tens of those big solar panels on a transport, but you could put hundreds of ultra-thin flexible ones on a transport and supply half the army," he said.

Most experts have estimated that flexible, solar cell technology for consumers was about a decade away, but Carroll said the new breakthrough at Wake Forest and NMSU means that consumers could be using this technology in the next five years.

Using a set of polymer coatings, researchers at Wake Forest constructed a nanophase within the polymer called a "mesostructure." The "mesostructure" changes the properties of the plastic and makes it better for collecting light. The researchers also removed the current from the polymer coating, Carroll said.

A test system at Wake Forest's nanotechnology center was used to simulate the sun, Carroll said, and the simulated spectrum was precisely measured and shot onto the organic solar cell, which appeared as a thin coat of paint. Devices at the center have registered almost 6 percent efficiency.

This breakthrough was announced in October at the Santa Fe Workshop on Nanoengineered Materials and Macro-Molecular Technologies, which was sponsored by Wake Forest's nanotechnology center.

Nanotechnology kills cancer cells

Nanotechnology has been harnessed to kill cancer cells without harming healthy tissue.

The technique works by inserting microscopic synthetic rods called carbon nanotubules into cancer cells.

When the rods are exposed to near-infra red light from a laser they heat up, killing the cell, while cells without rods are left unscathed.

Details of the Stanford University work are published by Proceedings of the National Academy of Sciences. Researcher Dr Hongjie Dai said: "One of the longstanding problems in medicine is how to cure cancer without harming normal body tissue.

"Standard chemotherapy destroys cancer cells and normal cells alike. "That's why patients often lose their hair and suffer numerous other side effects. "For us, the Holy Grail would be finding a way to selectively kill cancer cells and not damage healthy ones."

Many in cell

The carbon nanotubules used by the Stanford team are only half the width of a DNA molecule, and thousands can easily fit inside a typical cell.

Under normal circumstances near-infra red light passes through the body harmlessly.

But the Stanford team found that if they placed a solution of carbon nanotubules under a near-infra red laser beam, the solution heated up to about 70C in two minutes.

They then placed the tubules inside cells, and found they were quickly destroyed by the heat generated by the laser beam.

Dr Dai said: "It's actually quite simple and amazing. We're using an intrinsic property of nanotubes to develop a weapon that kills cancer."

The next step was to find a way to introduce the nantubules into cancer cells, but not healthy cells.

The researchers did this by taking advantage of the fact that, unlike normal cells, the surface of cancer cells is covered with receptors for a vitamin known as folate.

They coated the nanotubules with folate molecules, making it easy for them to pass into cancer cells, but unable to bind with their healthy cousins.

Exposure to the laser duly killed off the diseased cells, but left the healthy ones untouched.

Refined technique

The researchers believe it should be possible to refine the technique still further, for instance by attaching an antibody to a nanotubule to target a particular kind of cancer cell.

They have already started work on tailoring the technique to target lymphoma in mice.

Dr Emma Knight, of the charity Cancer Research UK, said: "Nanotechnology has a lot to offer biomedical science, and the results of this paper suggest yet another way in which it may help in the fight against cancer.

"However, this work is still at a very early stage. The researchers have shown that near-infra red light can cause nanotubes to produce heat that can kill cancer cells.

"But their work so far has focused on cells that have been grown in culture in the laboratory.

"Further research will be crucial to see whether these effects can be reproduced in the more complex environment of a tumour and, ultimately, the human body."

Particles Home In On Diseased Cells

Nanoparticles that reveal pancreatic cancer cells in mice have been developed in the laboratory of Ralph Weissleder at Harvard Medical School. Tools for early detection of pancreatic cancer in humans are lacking, and the nanoparticles may be a step toward better diagnosis of this deadly disease.

Weissleder and coworkers thought that decorating certain fluorescent magnetic nanoparticles—already being used clinically for magnetic resonance imaging—with small molecules would enable the nanoparticles to be taken up selectively by specific cells. The researchers have now shown that this idea works.

For example, nanoparticles studded with isatoic or 5-chloroisatoic anhydride are selectively taken up by pancreatic cancer cells in mice. Their presence inside those cells can be revealed by whole-body fluorescence or magnetic resonance imaging (*Nature Biotechnol.*, published Oct. 23.

Weissleder's approach is novel, says Robert S. Langer, professor of chemical engineering at Massachusetts Institute of Technology. "It could lead to a whole new way of targeting cells," he adds.

The work is a step toward giving nanomaterials more biological function—in this case, the ability to access specific cells or biological processes, Weissleder says. "One long-term goal is to use materials like these to detect diseases much earlier than currently possible," he adds.

The researchers prepared a library of 146 nanoparticles, each conjugated to an average of 60 copies of a small molecule. Screening revealed 14 with high affinity for pancreatic cancer cells. Among those, the particles decorated with isatoic acid or 5-chloroisatoic acid are poorly taken up by macrophages, which scavenge unmodified nanoparticles, and so were prepared on a large scale for the in vivo studies with mice.

THE JOINT PNNL/UW INSTITUTE FOR NANOSCIENCE & NANO LABS

Establishment

The University and Battelle established, effective as of January 1, 2001 the Joint Institute for Nanoscience to be governed, operated and managed in accordance with the terms and conditions set forth in the Joint Institutes Affiliation Agreement.

The Joint Institute for Nanoscience shall be located on the principal campus of the University in Seattle, Washington, but shall conduct activities at both the University and PNNL, as the Board and Steering Committee deem appropriate. For the purposes of this Institute, the following definitions shall apply: (i) "Nanoscience" shall mean the experimental and theoretical investigation of Nanoscale phenomena and shall also include Nanotechnology; (ii) "Nanoscale" shall mean dimensions between 0.1 and 300 nanometers, a nanometer being one-billionth (10^{-9}) of a meter; and (iii) "Nanotechnology" shall mean the application of Nanoscale phenomena.

Goals and Objectives

The goals and objectives of the Joint Institute for Nanoscience shall be to:

- strengthen existing collaborative research and development projects in Nanoscience at the University, PNNL, and any other mutually-agreed upon locations;
- coordinate educational and training opportunities in Nanoscience for University undergraduate students, graduate students, post-doctoral fellows, and others;
- assist in establishing faculty/staff development programs in Nanoscience at the University, PNNL, and any other mutually-agreed upon locations;
- assist in the development of multi-disciplinary programs in Nanoscience at the University, PNNL and any other mutually-agreed upon locations;
- coordinate short-course and conference offerings in Nanoscience;
- facilitate multi-institutional research efforts in Nanoscience, with special emphasis on partnerships among university, government, and industrial laboratories;
- establish a visiting scientist program that focuses initially on development of new instruments and techniques in Nanoscience, and subsequently on other areas of special interest;
- identify additional funding from public and private sector sources for the activities of the Joint Institute for Nanoscience; and
- make reasonable efforts to ensure the availability of facilities and equipment necessary to carry out the activities of the Joint Institute for Nanoscience activities.

Joint Institute for Nanoscience Steering Committee and Co-Directors

In accordance with the Agreement, the Steering Committee for the Joint Institute for Nanoscience shall consist of ten (10) members with the first-named undersigned signatories, or their designees, having the power to appoint on behalf of the University and Battelle, respectively, one-half of the membership. The term of office of a Steering Committee member shall be three (3) years and an appointing party may, at its sole discretion, remove or replace a member of the Steering Committee appointed by that party at any time or fill any vacancy in any position for which it has the right to make an appointment. Until otherwise changed by agreement of the parties, the parties shall each designate one person to serve as Co-Directors of the Joint Institute for Nanoscience. Both Co-Directors shall be members of the Steering Committee and alternate for terms of one (1) year in presiding over Steering Committee meetings. In the absence of the presiding officer, the other Co-Director shall preside over Steering Committee Meetings. The initial presiding officer shall be Battelle's appointee (currently Don Baer). The current UW co-director is Charles Campbell.

Steering Committee Responsibilities

The responsibilities of the Steering Committee of the Joint Institute for Nanoscience shall initially include the following, which may from time to time be modified by the Board:

- develop and propose to the Board, as requested, strategic and operational plans for the Joint Institute for Nanoscience;
- identify significant research areas and other programs in the field of Nanoscience offering significant potential;
- oversee research conducted by the Joint Institute for Nanoscience;
- prepare and submit, subject to the Board's approval, grant applications and other proposals to third parties for support of research in Nanoscience;
- periodically report to the Board the plans, progress, and problems of the Joint Institute for Nanoscience;
- develop, prepare and manage expenditure budgets as approved by the Board and parties hereto;
- conduct a search and make recommendations to the Board and the parties a candidate for the position of executive director of the Joint Institute for Nanoscience;
- make recommendations to the Board and the parties for appointments and reappointments on the Steering Committee;
- coordinate and schedule use of University and PNNL facilities for research and other activities of the Joint Institute for Nanoscience; and
- make recommendations to the Board and parties for participation by third parties in the activities of the Joint Institute for Nanoscience.

Financial Plan and Resource Commitments

The parties agreed to commit, on an annual basis, approximately $500,000 each for support of the Joint Institute for Nanoscience commencing. Most of this money will be awarded on an application basis, to support research collaborations between PNNL and UW. Award applicants can seek funding for graduate student, postdoc or faculty salary, tuitions for grad students, travel or housing expenses.

IMPORTANT LINKS

Laboratories

- Nanodesigners.net - Nanochemistry: synthesis, functionalization and assembly of nanoparticles
- NanoStructures Laboratory (NSL) - Nano-litography, Nano-electronics, Nano-magnetics
- IBM Almaden Laboratory - Atomic-scale manipulation and microscopy
- Zyvex company - Introduction and projects associated to nanotechnology
- Nasa nanotechnology's web site - Nanoelectronics
- The Institute of Nanotechnology - News in nanotechnology
- Nanotechnology Group, Institute of Robotics Zurich - Nanorobots and nanomanipulation
- The Bawendi Group Homepage - Semiconductors nanocrystals (quantum dots)
- The Alivisatos Group Homepage - Semiconductors nanocrystals (quantum dots) and biological applications
- The Brus Group Homepage - Semiconductors nanocrystals (quantum dots)
- The Colvin Group Homepage - Nanocrystals and porous solids
- The Morner Group Homepage - Single molecule Nanophotonics
- The Buhro Group Homepage - Nanofibers and nanowhiskers
- The Andres Group Homepage - Nanocrystals
- Nanoparticle Laboratory University of Melbourne - Nanocrystals
- The Penner Group Homepage - Metal or semiconductor nanocrystals
- Nanobiotechnology center at Cornell University - Nanobiotechnology
- The Vinayak P. Dravid Group Homepage - Magnetic nanoparticles and ceramic-matrix composites
- PNNL - Nanoscience, Nanoengineering and Nanotechnology

- Institute for Nanotechnology at Northwestern University - Nanofabrication
- The Centre for Nanoscale Science at the University of Liverpool - Nanotechnology and nanoparticle research
- Nano Science and Technology Center of Chinese Academy of Sciences

Companies: Nanotechnology and Nanomaterials products

- NanoLab Inc. - Carbon nanotubes
- Triton systems - Nanomaterials, coatings
- Scprod - Southern Clay Products
- Altairtechnologies - Oxide powders
- Nanox - Zinc Oxide ultrafine powders
- Qdots - Semiconductor nanoparticles
- Nextechmaterials - Ceramics
- Nanogen - DNA based microchip technology
- Argonide - Aluminium nanopowders
- Applied Sciences, Inc. - Advanced Materials (Carbon fibers and diamond films)
- Advanced nano-materials group - Various oxide nanomaterials
- Nanocor - Nanoclay technologies
- MicroPowder Solutions - Nanomaterials, advanced ceramics
- Cantilevers&Gratings for SPM
- Chengyin Technology (China) Co., Ltd. - nano-structured titanium dioxide in sunscreen, antimicrobial, anti-static and photocatalysis

Non profit organization

- Foresight Institute
- Réseau de Recherche en Micro et Nanotechnologies - In French
- The Institute of Nanotechnology (UK)

Government of United States

- National Nanotechnology Initiative
- Nanostructure Science and Technology
- Nanotechnology Database

Journals and News

- Nanotechnology
- Virtual Journal of Nanoscale Science & Technology
- Technology Review - MIT's Technology Magazine
- Smalltimes - News in Nanotechnology
- Nanotechnews.com - News in Nanotechnology
- nanotech-now.com - News in Nanotechnology
- Journal of Nanoscience and Nanotechnology
- Nanotechnology: It's a Small, Small, Small, Small World - By Ralph C. Merkle

Home pages

- Nicolas Duxin Home Page: Metal alloy nanoparticles
- Pramod H. Borse Home Page: University Of Pune (India) - Semiconductor quantum dots
- Ramon Aguado Home Page: Transport in quantum dots
- David Tomanek Home Page Nanoscale physics simulation
- Nicola Pinna Home Page: Semiconductor nanoparticles (all his thesis is online)

Specialists sites

- Nanotube page - All about nanotubes
- Quantum Well Structures and the Quantum Well Laser
- Raman Spectroscopy (French)
- Magnetic X-ray Circular Dichroism
- X-ray Diffraction
- ESCA - XPS
- An introduction to Squid
- Scanning probe microscopy

REFERENCES

Baird, D.; Nordmann, A. & Schummer, J. (eds.): 2004, *Discovering the Nanoscale*, Amsterdam: IOS Press.

Baker, R.T.K. Synthesis, properties and applications of graphite nanofibers. In *R&D status and trends*, ed. Siegel et al.

Bate, R., Frazier, G., Frensley, W., Reed., M., "An Overview of Nanoelectronics," *Texas Instruments Technical Journal*, July-August 1989, pp. 13-20.

Beck, J.S., J.C. Vartuli, W.J. Roth, M.E. Leonowicz, C.T. Kresge, K.D. Schmitt, C.T.-W. Chu, D.H. Olsen, E.W. Shepard, S.B. McCullen, J.B. Higgins, and J.L. Schlenker. 1992. *J. Am. Chem. Soc.* 114:10834.

Bensaude-Vincent, B.: 2001, 'The Construction of a Discipline: Materials Science in the United States', *Historical Studies in the Physical and Biological Sciences*, 31. 223-248.

Braun, T.; Schubert, A. & Zsindely, S.: 1997, 'Nanoscience and nanotechnology on the balance', *Scientometrics*, 38, 321-325.

Brinker, C.J. 1996. *Curr. Opin. Solid State Mater. Sci.* 1:798.

Brown, N.: 2004, 'Needle on the Real: Technoscience and Poetry at the Limits of Fabrication', in: N.K. Hayles (ed.), *Nanoculture: Implications of the New Technoscience*, Bristol, UK: Intellect Books, pp. 173-190.

Bueno, O.: 2004, 'The Drexler-Smalley Debated on Nanotechnology: Incommensurability at Work?', *Hyle: International Journal for Philosophy of Chemistry*, 10(2), 83-98.

Bueno, O.: 2004, 'Von Neumann, Self-Reproduction and the Constitution of Nanophenomena', in: D. Baird, A. Nordmann & J. Schummer (eds.), *Discovering the Nanoscale*, Amsterdam: IOS Press, pp. 101-115.

Crandall, B.C. and Lewis, J., *Nanotechnology: Research and Perspectives*, MIT Press, Cambridge, 1992.

Crow, M. & Sarewitz, D.: 2001, 'Nanotechnology and Societal Transformation', in: M.C. Roco and W.S. Bainbridge (eds.), *Societal Implications of Nanoscience and Nanotechnology*, Dordrecht: Kluwer, pp. 45-54.

Einsiedel, E.F.; Goldenberg, L.: 2004, 'Dwarfing the Social? Nanotechnology Lessons from the Biotechnology Front', *Bulletin of Science, Technology & Society*, 24, 28-33.

European Comission: *European Workshop on Social and Economic Research on Nanotechnologies and Nanosciences, Brussels, 14-15 April 2004* [www.stage-research.net/STAGE/PAGES/Nano.html].

European Commission (Community Health and Consumer Protection): 2004, *Nanotechnologies: A Preliminary Risk Analysis on theBasis of a Workshop, Brussels, 1-2 March 2004* [www.europa.eu.int/comm/health/ ph_risk/documents/ev_20040301_en.pdf].

European Commission: 2004, *Converging Technologies: Shaping the Future of European Societies (Report of the High Level Expert Group "Foresighting the New Technology Wave")*, Brussels: European Commission Research [http://europa.eu.int/comm/research/conferences/2004/ntw/index_en.html].

F. Buot, "Mesoscopic Physics and Nanoelectronics: Nanoscience and Nanotechnology," *Physics Reports*, pp.73-174, 1993.

F. Capasso and S. Datta, "Quantum Electron Devices," *Physics Today*, pp.74-82, February 1990.

Feynman, R., "There's Plenty of Room at the Bottom: An invitation to Enter a New Field of Physics," Talk at the Annual Meeting of the American Physical Society, 29 December 1959. Reprinted in Appendix B of Crandall and Lewis. See citation above.

Fielder, F.A.; Reynolds, G.H.: 1994, 'Legal Problems of Nanotechnology: An Overview', *Southern California Interdisciplinary Law Journal*, 3, 593-629.

Fleischer, Th.; Decker, M. & Fiedeler, U. (eds.): 2004, *Große Aufmerksamkeit für kleine Welten ' Nanotechnologie und ihre Folgen*, special issue of *Technikfolgenabschätzung ' Theorie und Praxis*, 13 (2), 5-85 [www.itas.fzk.de/tatup/042/inhalt.htm].

Fogelberg, H. & Glimell, H.: 2003, 'Molecular Matters: In Search of the Real Stuff', in: H. Fogelberg & H. Glimell, *Bringing Visibility to the Invisible: Towards A Social Understanding of Nanotechnology*, Göteborg: Göteborg University, pp. 5-32 [www.sts.gu.se/publications/STS_report_6.pdf].

Fogelberg, H. & Glimell, H.: 2003, *Bringing Visibility to the Invisible: Towards A Social Understanding of Nanotechnology*, Göteborg: Göteborg University [www.sts.gu.se/publications/STS_report_6.pdf].

Fogelberg, H.: 2003, 'The Grand Politics of Technoscience: Contextualizing Nanotechnology', in: H. Fogelberg & H. Glimell, *Bringing Visibility to the Invisible: Towards A Social Understanding of Nanotechnology*, Göteborg: Göteborg University, pp. 33-54 [www.sts.gu.se/publications/STS_report_6.pdf].

Fogelberg, H.: 2003, 'The Material Culture of Nanotechnology', in: H. Fogelberg & H. Glimell, *Bringing Visibility to the Invisible: Towards A Social Understanding of Nanotechnology*, Göteborg: Göteborg University, pp. 99-114.

Frazier, G., "An Ideology For Nanoelectronics," in *Concurrent Computations: Algorithms, Architecture, and Technology*, Plenum Press, New York, 1988.

Goldhaber-Gordon, D., Montemerlo, M.S., Love, J.C., Opiteck, G.J., and Ellenbogen, J.C., "Overview of Nanoelectronic Devices," submitted to the *Proceedings of the IEEE*, February 1997. For more information, please send e-mail to nanotech@mitre.org.

Gorman, M.E.; Groves, J.F. & Shrager, J.: 2004, 'Societal Dimensions of Nanotechnology as a Trading Zone: Results from a Pilot Project', in: D. Baird, A. Nordmann & J. Schummer (eds.), *Discovering the Nanoscale*, Amsterdam: IOS Press, pp. 63-73.

Grinbaum, A. & Dupuy, J.-P.: 2004, 'Living with Uncertainty: Toward the Ongoing Normative Assessment of Nanotechnology', *Techne: Research in Philosophy and Technology*, 8(3) (forthcoming).

Grinbaum, A.: 2004, 'La condition de l'homme moderne et les nanotechnologies', in: G. Nivat (ed.), *Les limites de l´humain. 39èmes Rencontres Internationales de Genève*, L´Age d´Homme, Genève, p. 141.

Grunwald, A.: 2004, 'Ethische Aspekte der Nanotechnologie. Eine Felderkundung', *Technikfolgenabschätzung ' Theorie und Praxis*, 13 (2), 71-78.

Gupta, V.K. & Pangannaya, N.B.: 2000, 'Carbon nanotubes: bibliometric analysis of patents', *World Patent Information*, 22, 185-189.

Hansson, S.O.: 2004, 'Great Uncertainty about Small Things', *Techne: Research in Philosophy and Technology*, 8(3) (forthcoming).

Haruta, M. 1997. *Catalyst surveys of Japan* 1:61 and references therein.

Hayles, N.K. (ed.): 2004, *Nanoculture: Implications of the New Technoscience*, Bristol, UK: Intellect Books.

Hayles, N.K.: 2004, 'Connecting the Quantum Dots: Nanotechscience and Culture', in: N.K. Hayles (ed.), *Nanoculture: Implications of the New Technoscience*, Bristol, UK: Intellect Books, pp. 11-26.

Hemerén, G.: 2004, 'Nano Ethics Primer', in: European Commission (Community Health and Consumer Protection): *Nanotechnologies: A Preliminary Risk Analysis on theBasis of a Workshop, Brussels, 1-2 March 2004* (www.europa.eu.int/comm/health/ ph_risk/documents/ ev_20040301_en.pdf), pp. 95-102.

Hennig, J.: 2004, 'Changes in the Design of Scanning Tunneling Microscopic Images from 1980 to 1990', *Techne: Research in Philosophy and Technology*, 8(3) (forthcoming).

Hessenbruch, A.: 2004, 'Nanotechnology and the Negotiation of Novelty', in: D. Baird, A. Nordmann & J. Schummer (eds.), *Discovering the Nanoscale*, Amsterdam: IOS Press, pp. 135-144.

Hullmann, A. & Meyer, M.: 2003, 'Publications and Patents in Nanotechnology. An overview of previous studies and the state of the art', *Scientometrics*, 58 (3) 507-527.

Huo, Q., D.I. Margolese, U. Ciesla, P. Feng, T.E. Gier, P. Sieger, R. Leon, P.M. Petroff, F. Schuth, and G.D. Stucky. 1994. *Nature* 368:317.

Jena, P., S.N. Khanna, and B.K. Rao. 1996. In *Science and technology of atomically engineered materials*, ed. P. Jena. River Edge, NJ: World Scientific.

Johansson, M.: 2003, 'Plenty of room at the bottom: Towards an anthropology of nanoscience', *Anthropology Today*, 19 (No. 6), 3-6.

Johnson, A.: 2004, 'The End of Pure Science: Science Policy from Bayh-Dole to the NNI', in: D. Baird, A. Nordmann & J. Schummer (eds.), *Discovering the Nanoscale*, Amsterdam: IOS Press, pp. 217-230.

Kresge, C.T., M.E. Leonowicz, W.J. Roth, J.C. Vartuli, and J.S. Beck. 1992. *Nature* 359:710.

Kupperman, A., S. Nadimi, S. Oliver, G. Ozin, J. Garcés, and M. Olken. 1993. *Nature* 365:239.

Kuusi, O.; Meyer, M.: 2002, 'Technological generalizations and leitbilder ' the anticipation of technological opportunities', *Technological Forecasting & Social Change*, 69, 625-639.

Landon, B.: 2004, 'Less is More: Much Less is Much More: The Insistent Allure of Nanotechnology Narratives in Science Fiction', in: N.K. Hayles (ed.), *Nanoculture: Implications of the New Technoscience*, Bristol, UK: Intellect Books, pp. 131-146.

Laszlo, P.: 2004, 'Is There Life After Partington?', *Hyle: International Journal for Philosophy of Chemistry*, 10(2), 169-178.

Lenhard, J.: 2004, 'Nanoscience and the Janus-Faced Character of Simulations', in: D. Baird, A. Nordmann & J. Schummer (eds.), *Discovering the Nanoscale*, Amsterdam: IOS Press, pp. 93-100.

Lent, C.S., Tougaw, P.D., Porod, W., Bernstein, G.H., "Quantum Cellular Automata," *Nanotechnology*, Vol. 4, p. 49, 1993.

Lewak, S.E.: 2004, 'What's the Buzz? Tell Me What's A-Happening: Wonder, Nanotechnology, and Alice's Adventures in Wonderland', in: N.K. Hayles (ed.), *Nanoculture: Implications of the New Technoscience*, Bristol, UK: Intellect Books, pp. 201-.

Lide, D.R., ed. 1993-1994. *CRC Handbook of Chemistry and Physics*, 74th ed.

Lin-Easton, P.C.: 2001, 'It's Time for Environmentalists to Think Small ' Real Small: A Call for the Involvement of Environmental Lawyers in Developing Precautionary Policies Molecular Nanotechnology', *Georgetown International Law Review*, 14, 106-134.

López, J.: 2004, 'Bridging the Gaps: Science Fiction in Nanotechnology', *Hyle: International Journal for Philosophy of Chemistry*, 10(2), 129-152.

Lösch, A.: 2004, 'Nanomedicine and Space: Discursive Orders of Mediating Innovations', in: D. Baird, A. Nordmann & J. Schummer (eds.), *Discovering the Nanoscale*, Amsterdam: IOS Press, pp. 193-202.

Paschen, H.; Coenen, C.; Fleischer, T.; Grünwald, R.; Oertel, D. & Revermann, C.: 2004, *Nanotechnologie: Forschung, Entwicklung, Anwendung*, Berlin: Springer (*Nanotechnologie, TAB-Arbeitsbericht 92*, Berlin: Büro für Technikfolgen-Abschätzung beim Deutschen Bundestag).

Roberts, J.A.: 2004, 'Deciding the Future of Nanotechnologies: Legal Perspectives on Issues of Democracy and Technology', in: D. Baird, A. Nordmann & J. Schummer (eds.), *Discovering the Nanoscale*, Amsterdam: IOS Press, pp. 247-255.

Robinson, C.: 2004, 'Images in NanoScience/Technology', in: D. Baird, A. Nordmann & J. Schummer (eds.), *Discovering the Nanoscale*, Amsterdam: IOS Press, pp. 165-169.

Robison, W.L.: 2004, 'Nano-Ethics', in: D. Baird, A. Nordmann & J. Schummer (eds.), *Discovering the Nanoscale*, Amsterdam: IOS Press, pp. 285-300.

Roco, M.C. & Bainbridge, W.S. (eds.): 2001, *Societal implications of nanoscience and nanotechnology*, (Proceedings of a workshop organized by the National Science Foundation, September 28-29, 2000), Kluwer: Dordrecht [available online at http://itri.loyola.edu/nano/societalimpact/nanosi.pdf]

Roco, M.C. & Tomellini, R. (eds.): 2002, *Nanotechnology: Revolutionary Opportunities and Societal Implications, Workshop, Lecce (Italy), 31 January - 1 February 2002*, Luxemburg: Office for Official Publications of the European Communities, [200 pp.]

Roco, M.C.; Bainbridge, W.S. (eds.): 2002, *Converging Technologies for Improving Human Performance: Nanotechnology, Biotechnology, Information Technology and the Cognitive Science*, Arlington, VA: National Science Foundation.

Roher, H. 1993. *Jpn. J. Appl. Phys.* 32:1335.

Rohlfing, E.A., D.M. Cox, and A. Kaldor. 1984. *J. Chem. Phys.* 81:3846.

Rosen, A. 1998. A periodic table in three dimensions: A sightseeing tour in the nanometer world. In *Advances in quantum chemistry*. In press.

Ruthven, D.M., S. Farooq, K.S. Knaebel. 1994. *Pressure swing adsorption*. New York: VCH Publishers.

Sarewitz, D.; Woodhouse, E.: 2003, 'Small is Powerful', in: A. Lightman, D. Sarewitz & Chr. Desser, (eds.), *Living with the Genie: Essays on Technology and the Quest for Human Mastery*, Washington, DC: Island Press, pp. 63-83.

Schiemann, G.: 2004, 'Dissolution of the Nature-Technology Dichotomy? Perspectives on Nanotechnology from an Everyday Understanding of Nature', in: D. Baird, A. Nordmann & J. Schummer (eds.), *Discovering the Nanoscale*, Amsterdam: IOS Press, pp. 209-213.

Schmidt, J.C.: 2004, 'Unbounded Technologies: Working Through Technological Reductionism of Nanotechnology', in: D. Baird, A. Nordmann & J. Schummer (eds.), *Discovering the Nanoscale*, Amsterdam: IOS Press, pp. 35-50.

Schuler, E.: 2004, 'Perception of Risks and Nanotechnology', in: D. Baird, A. Nordmann & J. Schummer (eds.), *Discovering the Nanoscale*, Amsterdam: IOS Press, pp. 279-284.

Schummer, J.: 2001, 'Ethics of Chemical Synthesis', *Hyle: International Journal for Philosophy of Chemistry*, 7, 103-124 [online: www.hyle.org/journal/issues/7/schummer.htm].

Schummer, J.: 2003, 'The Notion of Nature in Chemistry', *Studies in History and Philosophy of Science*, 34 (2003), 705-736.

Schummer, J.: 2004, 'Naturverhältnisse in der modernen Wirkstoff-Forschung', in: K. Kornwachs (ed.), *Technik ' System ' Verantwortung*, Münster: LIT, pp. 629-638.

Schummer, J.: 2004, 'Why do Chemists Perform Experiments?', in: D. Sobczynska, P. Zeidler, E. Zielonacka-Lis (eds.), *Chemistry in the Philosophical Melting Pot*, Peter Lang (Frankfurt/M.), 2004, pp. 395-410.

Schummer, J.; Baird, D. (eds.): 2004-5, *Nanotech Challenges*, joint special issue of *Hyle: International Journal for Philosophy of Chemistry & Techne: Research in Philosophy and Technology* (forthcoming).

Schwarz, A.E.: 2004, 'Shrinking the 'Ecological Footprint' with NanoTechnoScience?', in: D. Baird, A. Nordmann & J. Schummer (eds.), *Discovering the Nanoscale*, Amsterdam: IOS Press, pp. 203-208.

Stix Gary., "Toward 'Point One'," *Scientific American*, February 1995, pp. 90-95.

Stuckless, J.T, D.E. Starr, D.J. Bald, C.T. Campbell. 1997. *J. Chem. Phys.* 107:5547.

Suchman, M.: 2001, 'Envisioning Life on the Nano-Frontier', in: M.C. Roco and W.S. Bainbridge (eds.), *Societal Implications of Nanoscience and Nanotechnology*, Dordrecht: Kluwer, pp. 211-216.

2

Nano Labs & Nanotechnology Research

Research Programs on Nanotechnology in the World

Scientific breakthroughs combined with recent research programs in the United States, Japan, and Europe, and various initiatives in Australia, Canada, China, Korea, Singapore, and Taiwan highlight the international interest in nanoscale science and technology. Definitions of nanotechnology vary somewhat from country to country. Nanotechnology as defined for the projects reviewed in this chapter arises from the exploitation of the novel and improved physical, chemical, mechanical, and biological properties, phenomena, and processes of systems that are intermediate in size between isolated atoms/molecules and bulk materials, where phenomena length and time scales become comparable to those of the structure. It implies the ability to generate and utilize structures, components, and devices with a size range from about 0.1 nm (atomic and molecular scale) to about 100 nm (or larger in some situations) by control at atomic, molecular, and macromolecular levels. Novel properties occur compared to bulk behavior because of the small structure size and short time scale of various processes. Nanotechnology's size range and particularly its new phenomena set it apart from the technologies referred to as microelectromechanical systems (MEMS) in the United States or microsystems technologies (MST) in Europe.

It is estimated that nanotechnology is presently at a level of development similar to that of computer/information technology in the 1950s. As indicated in the preceding chapters and as evidenced by the WTEC panel's research and observations during the course of this study, the development of nanoscale science and technology is expected by most scientists working in the field to have a broad and fundamental effect on many other technologies. This helps to explain the phenomenal levels of R&D activity worldwide. This chapter presents an overview of most of the significant nanotechnology research programs in the world. Where possible, a general picture is given of the funding levels of the programs, based on site interviews and publications.

TheUnited States

Various U.S. public and private funding agencies; large companies in chemical, computer, pharmaceutical, and other areas; as well as small and medium-size enterprises provide support for precompetitive research programs on nanotechnology. Most of the supported programs are evolving out of disciplinary research programs, and only some are identified as primarily dealing with nanotechnology. U.S. government agencies sponsored basic research in this area at a level estimated at about \$116 million in 1997 (Siegel et al. 1998), as shown in Table 8.1. The National Science Foundation (NSF) has the largest share of the U.S. government investment, with an expenditure of about \$65 million per year, or about 2.4% of its overall research investment in 1997. In 1998 it expanded its research support to functional nanostructures with an initiative in excess of \$13 million.

NSF activities in nanotechnology include research supported by the Advanced Materials and Processing Program; the Ultrafine Particle Engineering initiative dedicated to new concepts and fundamental research to generate nanoparticles at high rates; the National Nanofabrication User Network (NNUN); and Instrument Development for Nano-Science and Engineering (NANO-95) to advance atomic-scale measurements of molecules, clusters, nanoparticles, and nanostructured materials. A current activity is the initiative, Synthesis, Processing, and Utilization of Functional Nanostructures (NSF 98-20 1997).

In the United States, a number of large multinational corporations, small enterprises, and consortia are pursuing nanotechnology-related research and development activities. Dow, DuPont, Eastman Kodak, Hewlett-Packard (HP), Hughes Electronics, Lucent, Motorola, Texas Instruments, Xerox, and other multinationals have established specialized groups in their long-term research laboratories, where the total research expenditure for nanotechnology research is estimated to be comparable to the U.S. government funding. Computer and electronics companies allocate up to half of their long-term research resources to nanotechnology programs. HP spends 50% of long-term (over 5 years) research on nanotechnology (Williams 1998). Small business enterprises, such as Aerochem Research Laboratory, Nanodyne, Michigan Molecular Institute, and Particle Technology, Inc., have generated an innovative competitive environment in various technological areas, including dispersions, coatings, structural materials, filtration, nanoparticle manufacturing processes, and functional nanostructures (sensors, electronic devices, etc.). Small niches in the market as well as support from several U.S. government agencies through the Small Business for Innovative Research (SBIR) program have provided the nuclei for high-tech enterprises. The university-small business technology transfer (STTR) program at NSF is dedicated to nanotechnology in fiscal year 1999. Two semiconductor processing consortia, the Semiconductor Manufacturing and Technology Institute (Sematech) and the Semiconductor Research Corporation (SRC), are developing significant research activities on functional nanostructures on inorganic surfaces. A series of interdisciplinary centers with nanotechnology activities has been established in the last few years at many U.S. universities, creating a growing public research and education infrastructure for this field. Examples of such centers are

- Brown University, Material Research Science and Engineering Center
- Rice University, Richard Smalley's Center for Nanoscale Science and Technology (CNST)
- University of California–Berkeley, nanoelectronics facilities
- University of Illinois at Urbana, the Engineering Research Center on Microelectronics in collaboration with the Beckman Institute, a private foundation
- University of North Carolina
- University of Texas–Austin
- Rensselaer Polytechnic Institute
- University of Washington (focus on nanobiotechnology)
- University of Wisconsin at Madison (focus on nanostructured materials)

NNUN, mentioned above, is an interuniversity effort supported by NSF at five universities: Cornell, Stanford, University of California–Santa Barbara (UCSB), Penn State, and Howard. It has focused on nanoelectronics, optoelectronics, electromechanical systems, and biotechnology. The Center for Quantized Electronic Structures (QUEST) at UCSB is a national facility developing expertise on underlying physics and chemistry aspects. Hundreds of graduate students have completed their education in connection with these centers in the last few years.

Current interest in nanotechnology in the United States is broad-based and generally spread into small groups. The research themes receiving the most attention include

1. metallic and ceramic nanostructured materials with engineered properties
2. molecular manipulation of polymeric macromolecules
3. chemistry self-assembling techniques of "soft" nanostructures
4. thermal spray processing and chemistry-based techniques for nanostructured coatings
5. nanofabrication of electronic products and sensors
6. nanostructured materials for energy-related processes such as catalysts and soft magnets
7. nanomachining
8. miniaturization of spacecraft systems

In addition, neural communication and chip technologies are being investigated for biochemical applications; metrology has been developed for thermal and mechanical properties, magnetism, micromagnetic modeling, and thermodynamics of nanostructures; modeling at the atomistic level has been established as a computational tool; and nanoprobes have been constructed to study material structures and devices with nanometer length scale accuracy and picosecond time resolution. While generation of nanostructures under controlled conditions by building up from atoms and molecules is the most promising approach, materials restructuring and scaling-down approaches will continue. Exploratory research includes tools of quantum control and atom manipulation, computer design of hierarchically structured materials (e.g., Olson 1997), artificially structured molecules, combination of organic and inorganic nanostructures, biomimetics, nanoscale robotics, encoding and utilization of information by biological structures, DNA computing, interacting textiles, and chemical and bioagent detectors.

Commercially viable technologies are already in place in the United States for some ceramic, metallic, and polymeric nanoparticles, nanostructured alloys, colorants and cosmetics, electronic components such as those for media recording, and hard-disk reading, to name a few. The time interval from discovery to technological application varies greatly. For instance, it took several years from the basic research discovery of the giant magnetoresistance (GMR) phenomenon in nanocrystalline materials (Berkowitz et al. 1992) to industry domination by the corresponding technology by 1997. GMR technology has now completely replaced the old technologies for computer disk heads, the critical components in hard disk drives, for which there is a $20+ billion market (Williams 1998). All disk heads currently manufactured by IBM and HP are based on this discovery. In another example, nanolayers with selective optical barriers are used at Kodak in more than 90% of graphics black and white film (Mendel 1997) and for various optical and infrared filters, which constitute a multibilliondollar business. Other current applications of nanotechnology are hard coatings, chemical and biodetectors, drug delivery systems via nanoparticles, chemical-mechanical polishing with nanoparticle slurries in the electronics industry, and advanced laser technology. Several nanoparticle synthesis processes developed their scientific bases decades ago, but most processes are still developing their scientific bases (Roco 1998). Most of the technology base development for nanoparticle work is in an embryonic phase, and industry alone cannot sustain the research effort required for establishing the scientific and technological infrastructure. This is the role of government (e.g., NSF and NIH) and private agency (e.g., Beckman Institute) support for fundamental research.

Nanotechnology research in the United States has been developed in open competition with other research topics within various disciplines. This is one of the reasons that the U.S. research efforts in nanotechnology are relatively fragmented and partially overlapping among disciplines, areas of relevance, and sources of funding. This situation has advantages in establishing competitive paths in the emerging nanotechnology field and in promoting innovative ideas; it also has some disadvantages for developing system applications. An interagency coordinating "Group on Nanotechnology" targets some improvement of the current situation. The group was established in 1997 with participants from twelve government funding/research agencies to enhance communication and develop partnerships among practicing nanoscience professionals.

Canada

Canada's National Research Council supports nanotechnology through the Institute for Microstructural Science, which has the mission to interact with industry and universities to develop the infrastructure for information technology. The main project, the Semiconductor Nanostructure Project, was initiated in 1990. It provides support for fundamental research at a series of universities, including Queen's, Carleton, and Ottawa Universities.

Japan

The term "nanotechnology" is frequently used in Japan specifically to describe the construction of nanostructures on semiconductors/inorganic substrates for future electronic and computer technologies, and to describe the development of equipment for measurement at nanometer level (Sienko 1998). There are, however, Japanese programs in a number of other areas related to nanotechnology in the broader definition used in this report.

Government agencies and large corporations are the main sources of funding for nanotechnology in Japan; small and medium-size companies play only a minor role. Research activities are generally grouped in relatively large industrial, government, and academic laboratories. The three main government organizations sponsoring nanotechnology in Japan are the Ministry of International Trade and Industry (MITI), the Science and Technology Agency (STA), and Monbusho (the Ministry of Education, Science, Sports, and Culture). Funding for nanotechnology research should be viewed in the context of an overall increased level of support for basic research in Japan since 1995 as a result of Japan's Science and Technology Basic Law No. 130 (effective November 15, 1995), even if the law has not been fully implemented. The data presented below are based on information received from Japanese colleagues during the WTEC visit in July 1997. All budgets are for the fiscal year 1996 (1 April 1996 to 31 March 1997) and assume an exchange rate of ¥115 = $1, unless otherwise stated. The first five-year program on ultrafine particles started in 1981 under the Exploratory Research for Advanced Technologies (ERATO) program; an overview of the results of that program was published in 1991 (Uyeda 1991).

It is estimated that the Agency of Industrial Science and Technology (AIST) within MITI had a budget of approximately $60 million per year for nanotechnology in 1996/97 (roughly 2.2% of the AIST budget). The National Institute for Advancement of Interdisciplinary Research (NAIR) hosts three AIST projects:

1. Joint Research Center for Atom Technology (JRCAT), with a ten-year

budget of about $220 million for 1992-2001 ($25 million per year in 1996)

2. Research on Cluster Science program, with about $10 million for the interval 1992-1997
3. Research on Bionic Design program, with $10 million for 1992-1997, about half on nanotechnology

Other efforts supported to various degrees by MITI include the following:

- the Electrotechnical Laboratory in Tsukuba, which allocates about 17% (or $17 million per year) of its efforts on advanced nanotechnology projects
- the Quantum Functional Devices Program, funded at about $64 million for 1991-2001 (about $6.4 million in 1996)
- the Osaka National Research Institute and the National Industrial Research Institute of Nagoya, which each spend in the range of $2.5-3 million per year for nanotechnology
- the Association of Super-Advanced Electronics Technologies (ASET), a relatively new MITI-sponsored consortium with partial interest in nanotechnology; it has similarities with the U.S. Ultra Electronics program of DARPA

It is estimated that STA investment in nanotechnology research was about $35 million in 1996, mainly within five organizations:

1. Institute of Physical and Chemical Research (RIKEN), where nanotechnology is included in the Frontier Materials Research initiative
2. National Research Institute for Metals (NRIM)
3. National Institute for Research in Inorganic Materials (NIRIM)
4. Japan Science and Technology Corporation (JST—formerly called JRDC), which manages the ERATO program, including four nanotechnology-related projects, each with total budgets of $13-18 million for five years:
   - Quantum Wave Project (1988-1993)
   - Atomcraft Project (1989-1994)
   - Electron Wavefront Project (1989-1994)
   - Quantum Fluctuation Project (1993-1998)

Monbusho supports nanotechnology programs at universities and national research institutes, as well as via the Japan Society for Promotion of Science (JSPS). The most active programs are those at Tokyo University, Kyoto University, Tokyo Institute of Technology, Tohoku University, and Osaka University. The Institute of Molecular

Science and the Exploratory Research on Novel Artificial Materials and Substances program promote new research ideas for next-generation industries (5-year university-industry research projects). The "Research for the Future" initiative sponsored by JSPS has a program on "nanostructurally controlled spin-dependent quantum effects" (1996-2001) at Tohoku University. Monbusho's funding contribution to nanotechnology programs is estimated at ~ $25 million.

In total, MITI, STA, and Monbusho allocated ~ $120 million for nanotechnology in 1996.

Large companies also drive nanotechnology research in Japan. Important research efforts are at six institutions: Hitachi (Central R&D Laboratories, where nanotechnology is ~ 25% of long-term research)—Hitachi has ~ $70 billion per year in sales; NEC (Fundamental Research Laboratories, where nanotechnology is estimated to be ~ 50% of the precompetitive research)—NEC has ~ $40 billion per year in sales; NTT (Atsugi Lab); Fujitsu (Quantum Electron Devices Lab); Sony; and Fuji Photo Film Co. An allocation of 10% of sales for research and development is customary in these companies, with ~ 10% of this for long-term research. Some Japanese nanostructured products already have considerable market impact. Nihon Shinku Gijutsu (ULVAC) produces over $4 million per year in sales of particles for electronics, optics, and arts. Also, there are in Japan, at the United States, private consortia making an increased contribution to nanotechnology R&D:

- The Semiconductor Industry Research Institute of Japan (SIRI), established in 1994, focuses on long-term research with partial government funding
- Semiconductor Leading Edge Technologies, Inc. (SELETE), established by ten large Japanese semiconductor companies in 1996, focuses on applied research and development with an estimated budget of $60 million in 1997
- Semiconductor Technology Academic Research Center (STARC) promotes industry-university interactions

Strengthening of the nanotechnology research infrastructure in the last years has been fueled by both the overall increase of government funding for basic research and by larger numbers of academic and industry researchers choosing nanotechnology as their primary field of research. Potential industrial applications provide a strong stimulus. A systems approach has been adopted in most laboratory projects, including multiple characterization methods and processing techniques. A special Japanese research strength is instrumentation development. The university-industry interaction is stimulated by the new MITI projects awarded to universities in the last few years that encourage temporary hiring of research personnel from industry. Other issues currently being addressed are more extensive use of peer review, promoting personnel mobility and intellectual independence, rewarding researchers for patents, promoting interdisciplinary and international interactions, and better use of the physical infrastructure.

China

Nanoscience and nanotechnology have received increased attention in China since the mid-1980s. Approximately 3,000 researchers there now contribute to this field (Bai 1996). The ten-year "Climbing Project on Nanometer Science" (1990-1999) and a series of advanced materials research projects are the core activities. The Chinese Academy of Sciences sponsors relatively large groups, while the China National Science Foundation (CNSF) provides support mainly for individual research projects. Areas of strength are development of nanoprobes and manufacturing processes using nanotubes. The Chinese Physics Society and the Chinese Society of Particuology are societies involved in the dissemination of nanotechnology research.

India

India's main research activities are on nanostructured materials and electronic devices (Sikka 1995). These involve a combination of research institutions (the Central Electronics Engineering Research Institute in Pilani, the Space Application Center in Ahmedabad, and others), funding organizations (the Centre for Development of Materials in Pune, the Indian Institute of Science in Bangalore, and others), and industry.

Taiwan

Taiwan's major nanotechnology research effort is conducted in the area of miniaturization of electronic circuits. The research is conducted in academic institutions and at the Industrial Technology Research Institute. Government funds for fundamental research are channeled via the National Science Council.

South Korea

A national research focus on nanotechnology was established in Korea in 1995. The Electronics and Telecommunications Research Institute (ETRI) in Taejon, Korea's science city, targets advanced technologies for information and computer infrastructures, with a focus on nanotechnology (ATIP 1998). The emphasis is on nanoscale semiconductor devices and particularly on semiconductor quantum nanostructures and device applications (lasers, modulators, switches and logical devices, resonant tunneling devices, self-assembled nanosize dots, single-electron transistors, and quantum wires).

Singapore

Nanotechnology research received a considerable boost in Singapore by the initiation of a national program in this area in 1995.

Australia

The National Research Council (NRC) of Australia has sponsored R&D in nanotechnology since 1993 (ASTC 1993). Research groups work on synthesis of nanoparticles for membranes and catalysts (University of New South Wales), nanofiltration (UNESCO Center for Membrane Science and Technology), and use of nanoparticles in processing minerals for special products (the Advanced Mineral Products Research Center at the University of Melbourne). AWA Electronics in Homebush has the largest industrial research facility for nanoelectronics in Australia.

European Community (EC)

The term "nanotechnology" is frequently defined in Europe as "the direct control of atoms and molecules" for materials and devices. A more specific definition from H. Rohrer (1997) is a "one-to-one relationship between a nano-object or nano-part of an object and another nano-, micro- or bulk object." The nanotechnology field as defined in this WTEC report includes these aspects, with the clarification that only the specific, distinctive properties and phenomena manifesting at length scales between individual atoms/molecules and bulk behavior are considered.

There are a combination of national programs, collaborative European (mostly EC) networks, and large corporations that fund nanotechnology research in Europe. Multinational European programs include the following:

1. The ESPRIT Advanced Research Initiative in Microelectronics and the BRITE/EURAM projects on materials science in the EC are partially dedicated to nanotechnology.
2. The PHANTOMS (Physics and Technology of Mesoscale Systems) program is a network created in 1992 with about 40 members to stimulate nanoelectronics, nanofabrication, optoelectronics, and electronic switching. Its coordinating center is at the IMEC Center for Microelectronics in Leuven, Belgium.
3. The European Science Foundation has sponsored a network since 1995 for Vapor-phase Synthesis and Processing of Nanoparticle Materials (NANO) in order to promote bridges between the aerosol and materials science communities working on nanoparticles. The NANO network includes 18 research centers and is codirected by Duisburg University and Delft University of Technology.
4. The European Consortium on NanoMaterials (ECNM) was formed in 1996, with its coordinating center in Lausanne, Switzerland. This group aims at fundamental research to solve technological problems for nanomaterials and at improved communication between researchers and industry.
5. NEOME (Network for Excellence on Organic Materials for Electronics) has had some programs related to nanotechnology since 1992.
6. The European Society for Precision Engineering and Nanotechnology

(EUSPEN) was designed in 1997 with participation from industry and universities from six EC countries.

7. The Joint Research Center Nanostructured Materials Network, established in 1996, has its center in Ispra, Italy.

It is expected that the European Framework V will introduce additional programs on nanotechnology, particularly by adding a new dimension in nanobiology in the next four-year plan.

Germany

The Federal Ministry of Education, Science, Research, and Technology (BMBF) in Germany provides substantial national support for nanotechnology. The Fraunhofer Institutes, Max Planck Institutes, and several universities have formed centers of excellence in the field. It is estimated that in 1997 BMBF supported programs on nanotechnology with a budget of approximately $50 million per year. Two of the largest upcoming projects are "CESAR," a $50 million science center in Bonn equally sponsored by the state and federal governments with about one-third of its research dedicated to nanoscience, and a new institute for carbon-reinforced materials near Karlsruhe ($4 million over 3 years, 1998-2001). BMBF is establishing five "centers of competence in nanotechnology" in Germany starting in 1998, with topics ranging from molecular architecture to ultraprecision manufacturing.

U.K.

A network program (LINK Nanotechnology Programme) was launched in the United Kingdom in 1988 with an annual budget of about $2 million per year. The Engineering and Physical Sciences Research Council (EPSRC) is funding materials science projects related to nanotechnology with a total value of about $7 million for a five-year interval (1994-1999). About $1 million is specifically earmarked for nanoparticle research. The National Physical Laboratory established a forum called the National Initiative on Nanotechnology (NION) for promoting nanotechnology in universities, industry, and government laboratories.

France

The Centre National de la Récherche Scientifique (CNRS) has developed research programs on nanoparticles and nanostructured materials at about 40 physics laboratories and 20 chemistry laboratories in France. Synthesis methods include molecular beam and cluster deposition, lithography, electrochemistry, soft chemistry, and biosynthesis. Nanotechnology activity has grown within a wide variety of research groups, including ones focused on molecular electronics, large gap semiconductors and nanomagnetism, catalysts, nanofilters, therapy problems, agrochemistry, and even cements for ductile nanoconcretes. It is estimated that CNRS spends about 2% of its budget and dedicates 500 researchers in 60 laboratories (or about $40 million per year) on projects related to nanoscience and nanotechnology. Companies collaborating to research and produce nanomaterials include Thompson, St. Gobain, Rhône Poulenc, Air Liquide, and IEMN. Also, there is the "French Club Nanotechnologie," aimed at promoting interactions in this field in France.

Sweden

The estimated total expenditure for research on nanotechnology in Sweden is ~ $10 million per year. There are four materials research consortia involved in this field: 1. Ångström Consortium in Uppsala, with a budget of ~ $0.8 million per year in 1998 for surface nanocoatings 2. Nanometer Structures Consortium in Lund with a budget of ~ $3.5 million per year partially supported by ESPRIT (~ $1 million per year) 3. Cluster-based and Ultrafine Particle Materials in the University of Uppsala and Royal Institute of Technology with a budget of ~ $0.4 million per year in 1998 4. Brinell Center at the Royal Institute of Technology

Switzerland

There is a Swiss national program on nanotechnology with a special strength in instrumentation. The most advanced research centers are focused on nanoprobes and molecule manipulation on surfaces (IBM Research Laboratory in Zürich), devices and sensors (Paul Scherrer Institute), nanoelectronics (ETH Zürich), and self-assembling on surfaces in patterns determined by the substrate or template (L'École Polytechnique Fédérale de Lausanne). See the site reports on all of these institutions.

The Netherlands

The most active research centers in the Netherlands are the DIMES institute at Delft University of Technology, which receives one-third support from industry, and the Philips Research Institute in Eindhoven, which researches self-assembling monolayers and patterning on metallic and silicon surfaces. The SST Netherlands Study Center for Technology Trends is completing a study on nanotechnology and aims at promoting increased funding and research networking in the Netherlands (ten Wolte 1997).

Finland

The Academy of Finland and the Finnish Technology Development Center began a three-year nanotechnology program in 1997. The program involves sixteen projects with funding of $9 million for a three-year period (1997-1999) for nanobiology, functional nanostructures, nanoelectronics, and other areas. Research on actuators and sensors is the Finnish area of strength.

Belgium and Spain

Since about 1993 both Belgium and Spain have established nanotechnology programs, centers of excellence, and university-industry interactions.

Multinational Efforts

Large multinational companies with significant nanotechnology research activities in Europe include IBM (Zürich), Philips, Siemens, Bayer, and Hitachi. Degussa Co., with headquarters in Germany, is a commercial supplier since 1940 of microparticles and, now, nanoparticles.

Western Europe has a variety of approaches to funding research on nanotechnology. These are discussed in detail in other studies. IPTS (the Science and Technology Forecast Institute) has conducted a study on nanotechnology research in the EC (Malsh 1997). Other European studies published recently include those by VDI (1996), UNIDO (1997), the U.K. Parliamentary Office of Science and Technology (1997), and NANO network (Fissan and Schoonman 1997). The overall expenditure for nanotechnology research within the EC was estimated in 1997 to be over $128 million per year.

Russia and Other FSU Countries

Support for generation of nanoparticles and nanostructured materials has a tradition in Russia and other countries of the former Soviet Union (FSU) dating back to the mid-1970s; before 1990 an important part of this support was connected to defense research. The first public paper concerning the special properties of nanostructures was published in Russia in 1976. In 1979 the Council of the Academy of Sciences created a section on "Ultra- Dispersed Systems." Research strengths are in the areas of preparation processes of nanostructured materials and in several basic scientific aspects. Metallurgical research for special metals, including those with nanocrystalline structures, has received particular attention; research for nanodevices has been relatively less developed. Due to funding limitations, characterization and utilization of nanoparticles and nanostructured materials requiring costly equipment are less advanced than processing.

Russian government funds are allocated mainly for research personnel and less for infrastructure. Funding for nanotechnology is channeled via the Ministry of Science and Technology, the Russian Foundation for Fundamental Research, the Academy of Sciences, the Ministry of Higher Education, and other ministries with specific targets. The Ministry of Higher Education has relatively little research funding. Overall, 2.8% of the civilian budget in Russia in 1997 was planned for allocation to science. There is no centralized program on nanotechnology; however, there are components in specific institutional programs. Currently, about 20% of science research in Russia is funded via international organizations. The significant level of interest in the FSU can be identified by the relatively large participation at a series of Russian conferences on nanotechnology, the first in 1984 (First USSR Conference on Physics and Chemistry of Ultradispersed Systems), a second in 1989, and a third in 1993.

The Ministry of Science and Technology contributes to nanotechnology through several of its specific programs related to solid-state physics, surface science, fullerenes and nanostructures, and particularly "electronic and optical properties of nanostructures." This last program involves a network of scientific centers: the Ioffe Institute in St. Petersburg, Lebedev Institute in Moscow, Moscow State University, Novgograd Institute of Microstructures, Novosibirsk Institute of Semiconductor Physics, and others. This research network has an annual meeting on nanostructures, physics, and technology, and has developed interactions with the PHANTOMS network in the EC. The U.S. Civilian Research and Development Foundation has provided research funds in the FSU for several projects related to nanotechnology, including "Highly Non-Equilibrium States and Processes in Nanomaterials" at the Ioffe Institute (1996-1998).

Russian government and international organizations are the primary research sponsors for nanotechnology in Russia. However, laboratories and companies privatized in the last few years, such as the Delta Research Institute in Moscow, are under development. With a relatively lower base in characterization and advanced computing, the research focus is on advanced processing and continuum modeling. Research strengths are in the fields of physico-chemistry, nanostructured materials, nanoparticle generation and processing methods, and applications for hard materials, purification, and the oil industry, and biologically active systems (Siegel et al. n.d.). There are related programs in Ukraine, Belarus, and Georgia, mostly under the direction of the respective academies of sciences in these countries, that are dedicated to crystalline nanostructures and advanced structural and nanoelectronic materials. Several innovative processes, such as diamond powder production by detonation synthesis at SINTA in Belarus, are not well known abroad.

Nanotechnology in the United States, Japan, and Western Europe is making progress in developing a suitable research infrastructure. The promise of nanotechnology is being realized through the confluence of advances in two fields: (1) scientific discovery that has enabled the atomic, molecular, and supramolecular control of material building blocks, and (2) manufacturing that provides the means to assemble and utilize these tailored building blocks for new processes and devices in a wide variety of applications. Technology programs cannot be developed without strong supporting science programs because of the scale and complexity of the nanosystems. The overlapping of discipline-oriented research with nanotechnology-targeted programs seems appropriate at this point in time. Highly interdisciplinary and multiapplication nanotechnology provides generic approaches that enable advances in other technologies, from dispersions, catalysts, and electronics to biomedicine. Essential trends include the following:

- learning from nature (including templating, self-assembly, multifunctionality)
- building up functional nanostructures from molecules
- convergence of miniaturization and assembly techniques
- novel materials by design
- use of hierarchical/adaptive simulations

A characteristic of discovery in nanotechnology is the potential for revolutionary steps. The question "what if?" is progressively replaced by "at what cost?" The road from basic research to applications may vary from a few months to decades. Research and development is expensive, and the field needs support from related areas. The R&D environment should favor multiapplication and international partnerships. Based on the data for 1996 and 1997 collected during this WTEC study, 1997 government expenditures for nanotechnology research were at similar absolute levels in the United States, Japan, and Western Europe. (Estimated OECD data for 1997 indicated GDPs of $4.49 billion for Japan, $7.76 billion for the United States, and $7.00 billion for Western Europe.) The largest funding opportunities for nanotechnology are provided by NSF in the United States (approximately $65 million per year for fundamental research), by MITI in Japan (approximately $50 million per year for fundamental research and development), and by BMBF in Germany (approximately $50 million per year for fundamental and applied research). Large companies in areas such as dispersions, electronics, multimedia, and bioengineering contribute to research to a larger extent in Japan and the United States than in Europe.

While multinational companies are pursuing nanotechnology research activities in almost all developed countries, the presence of an active group of small and medium-size companies introducing new processes to the market is limited to the United States. In the United States, individual and small-group researchers as well as industrial and national laboratories for specialized topics have established a strong position in synthesis and assembly of nanoscale building blocks and catalysts, and in polymeric and biological approaches to nanostructured materials. The Japanese large-group research institutes, and more recently academic laboratories, have made particular advances in nanodevices and nano-instrumentation. The European "mosaic" provides a diverse combination of university research, networks, and national laboratories with special performance in dispersion and coatings, nanobiotechnology, and nanoprobes.

With a relatively lower base in characterization and computing infrastructure, the research focus in Russia is on physico-chemistry phenomena, advanced processing, and continuum modeling. Interest and economic support, particularly for device-related research, is growing in China, Australia, India, Taiwan, Korea, and Singapore.

The pace of revolutionary discoveries that we are witnessing now in nanotechnology is expected to accelerate in the next decade worldwide. This will have a profound impact on existing and emerging technologies in almost all industry sectors, in conservation of materials and energy, in biomedicine, and in environmental sustainability.

Research Centers

- Institute for Molecular Manufacturing - nonprofit foundation carrying out research aimed at developing molecular manufacturing (molecular nanotechnology).

- Rice University - Center for Nanoscale Science and Technology - devoted to nurturing science and technology at the nanometer scale.
- Cornell NanoScale Science & Technology Facility (CNF) - subjects of research encompass physical sciences, life sciences, and engineering, particularly with inter-disciplinary emphasis.
- Cornell University - Nanobiotechnology Center (NBTC) - devoted to studying the convergence of nano-microfabrication and biosystems.
- University of Washington - Center for Nanotechnology University of Birmingham (United Kingdom) - Nanoscale Physics Research Laboratory - advancing the frontiers of the physics, chemistry, and technology of nanometre-scale structures, devices, and processes.
- University of Newcastle upon Tyne - Centre for Nanoscale Science and Technology - multidisciplinary research centre developing micro- and nano-devices for biomedical applications.
- University of California, Santa Barbara Nanotech - offers expertise in compound semiconductor-based device fabrication providing a full range of processes to the scientific and research communities.
- Center for Nanoscience and Nanotechnology - Georgia Institute of Technology - working to improve the human condition by providing fundamental and applicable knowledge to nanotechnology and its related fields of science.
- University of Michigan - Center for Biologic Nanotechnology - dedicated to the application of nano-molecular architecture to medicine.
- University of Notre Dame Nanofabrication Facility (NDNF) - provides information on the nanoelectronics fabrication facility.
- Arizona State University - Graduate Research Training Program in Biomolecular Nanotechnology - offers training in biomolecular nanotechnology, molecular electronics, and molecular devices.
- University of Nebraska-Lincoln - Center for Electro-Optics - specializes in optical/electromagnetic measurement and remote sensing technologies research.
- Schubert-Group: Macromolecular Chemistry and Nanoscience - Eindhoven University of Technology - devoted to the combinatorial materials research efforts.
- California Nanosystems Institute (CNSI) - R&D institute working on nanotechnologies and its commercial applications. Jointly established by University of California, Los Angeles, and University of California, Santa Barbara.

- Center for Nanotechnology - NASA Ames Research Center - focuses on experimental research and development in nano and bio technologies as well as computational nanotechnology, nanoelectronics, and optoelectronics.
- Institute for Soldier Nanotechnologies - intent to create a University Affiliated Research Center (UARC), with industry partners, to develop nanometer-scale science and technology solutions for the soldier.
- National Institute of Standards and Technology - Nanometer-Scale Metrology - research projects and measurement services in nanometer scale metrology

World-Wide Nanoelectronics Research

U.S. Government Labs

- Department of Defense (DOD) Laboratories

  DOD Labs conduct some of the most advanced and imaginative research in nanofabrication, nanoelectronics, and nanocomputing. Not all of this research is well represented on the World-Wide Web, however. THE Nanoelectronics & Nanocomputing Home Page has attempted to list some of this research, at least by title, on the Who's Who in Nanoelectronics page, as well as other pages. For further information on DOD Labs and an index of DOD technology departments and projects, see the TechTRANSIT Page. TechTRANSIT connects technology transfer resources and activities to meet the requirements of the Department of Defense Office of Technology Transition.

- Naval Research Laboratory (NRL)

  The Naval Research Laboratory (NRL) is the Navy's corporate research and development laboratory. NRL is the leader among Government Laboratories in nanotechnology and nanocomputing research. This includes very strong efforts in the application of proximal probes, modeling of nanosystems, novel approaches to nanofabrication, as well as other areas. However, only limited information about NRL's nanoelectronics activities is available on their WWW pages. For some information on nanoelectronics work at the NRL, see the NRL Electronics Science & Technology Division.

- Los Alamos National Laboratory (LANL).

  Investigators at LANL are deeply involved in the study of nanoelectronics and nanofabrication as part of the Theory, Modeling, and High Performance Computing core competency at the laboratory. However, only limited information about LANL's nanoelectronics activities is available on their WWW pages. Some of the LANL investigators are represented, however, on the Who's Who in Nanoelectronics page.

U.S. Universities

- UCLA Nanoelectronics Lab
- The University of Cincinnati Nanoelectronics Lab

  The Nanoelectronics Laboratory is engaged in research at (or around) the nanometer scale in semiconductors and other electronic materials. Specific research areas include focused ion beam processing, integrated optoelectronic devices and waveguides, plasma assisted etching, and deposition.
- Cornell University Nanofabrication Facility (CNF)

  The Cornell Nanofabrication Facility (CNF)has served the US research community for more than 15 years by providing nanofabrication services to over 500 research projects. Subject areas of projects completed are as diverse as Astronomy to Plant Pathology, as well as the more conventional Micro-electronics, Physics and Materials Research, and emerging fields of Integrated Optics and Micro-Electro-Mechanical Systems(MEMS).
- Massachusetts Institute of Technology (MIT) NanoStructures Laboratory

  The NanoStructures Laboratory (NSL) at MIT (formerly the Submicron Structures Laboratory) develops techniques for fabricating surface structures with feature sizes in the range from nanometers to micrometers, and uses these structures in a variety of research projects.
- University of Minnesota Nanostructure Lab

  The Nanostructure Lab's goal is to develop new nanotechnologies for fabricating structures and devices substantially smaller than current technology permits, and to explore innovative nanoscale electronic, optoelectronic, and magnetic storage devices. Research at the lab is headed by Stephen Chou.
- Notre Damè Microelectronics Lab

  The Microelectronics Laboratory is equipped to fabricate integrated circuits and devices in both silicon and compound semiconductors with geometries as small as 0.01 microns (10 nm).
- Penn State Electronic Materials and Processing Research Laboratory (EMPRL)
- Purdue University Nanoscale Physics Laboratory

  This Laboratory is a part of an interdisciplinary Center of Excellence in nanotechnology that has been developed at Purdue 'University as part of a University Research Initiative funded via the Army Research Office. Among the accomplishments of the interdisciplinary, interdepartmental

Purdue team is the fabrication of arrays of functioning nanometer-scale circuit elements and wires using techniques of chemical self-assembly.

- Stanford University Nanofabrication Facility

  The Stanford Nanofabrication Research Facility is located in the Center for Integrated Systems. This state-of-the-art research laboratory consists of over $10M worth of modern processing equipment in a 10,000 sq.ft. clean room, extensive computing and networking capability, established processes for silicon, gallium arsenide and other material, a highly skilled technical staff and some of the most advanced research projects in the world.

U.S. Government Sponsored Research and Development Programs

- Ultra Dense, Ultra Fast Computing Components/Nanoelectronics (ULTRA) Program

  Sponsored by the Defense Advanced Research Projects Agency, the goals of the ULTRA Nanoelectronics program are to explore and develop material, processing technologies, quantum and conventional devices and device architectures for a next generation of information processing systems. The ULTRA program seeks improved speed, density, and functionality beyond that which can be achieved by scaling transistors.

- Nanoelectronics Modeling (NEMO) Program

  The U.S. Government's Project NEMO is developing an advanced software tool for modeling nanoelectronic devices.

U.S. Government Sponsoring Agencies

- Defense Advanced Research Projects Agency (DARPA)

  The Defense Advanced Research Projects Agency (DARPA)is the advanced research and development organization for the U.S. Department of Defense. More information on DARPA's nanoelectronics work is available at the DARPA Electronics Technology Office(ETO)Home Page. The DARPA ULTRA Program (for the development of Ultra-dense and Ultra-fast computers) is the premier nanoelectronics research program in the world.

- Air Force Office of Scientific Research (AFOSR)

  The Air Force Office of Scientific Research (AFOSR) is charged with directing the Air Force's basic research program, which includes extensive efforts in next-generation electronics and related technologies.

- Army Research Office (ARO)

The Army Research Office sponsors a diverse set of research and development projects in nanoelectronics. ARO is led by Dr. Gerry Iafrate, who was a very early supporter and investigator in the fields of quantum electronics and nanoelectronics.

- Office of Naval Research (ONR)

  The Office of Naval Research (ONR) coordinates, executes, and promotes the science and technology programs of the United States Navy and Marine Corps through universities, government laboratories, and nonprofit organizations. The ONR Electronics Division has a nanoscale electronics program. The objective of this program is to study the electronic properties of solid state materials and develop ultra small devices for microelectronic applications.

- The National Science Foundation (NSF)

  NSF's mission is to promote the progress of science; to advance the national health, prosperity, and welfare; and to secure the national defense. NSF funds research at U.S. universities in nanoelectronics, and it also has considerable involvement in the funding of DNA/biomolecular nanocomputation.

Federally Funded Research and Development Centers (FFRDCs)

- The Aerospace Corporation

  Aerospace operates a Federally Funded Research and Development Center serving as an architect-engineer for advanced space systems. Aerospace has been concerned with developing and applying nanoelectronics and nanotechnology to ease the manufacture and, especially, to reduce the size and weight of spacecraft. This could have dramatic impact in reducing launch costs, as well as in improving reliability. Aerospace has worked with NASA to sponsor several meetings on these topics, and they have published technical reports and proceedings in connection with these activities. To learn more about nanoelectronics at The Aerospace Corporation, see the Web pages for their Electronic Systems Division.

- MIT Lincoln Laboratory

  Dr. Gerhard Sollner of the MIT Lincoln Lab is a well-known investigator at the forefront of nanoelectronics research.

- The MITRE Corporation

  Nanoelectronics research at The MITRE Corporation is performed by the MITRE Nanosystems Group. The MITRE Corporation is a Federally Funded Research and Development Center (FFRDC) for systems engineering and integration serving in the public interest. MITRE applies leading-edge technology to some of the most important systems in the world, including U.S. Military command and control systems and air traffic control systems, as well as similar systems around the world.

- The RAND Corporation

  RAND is a nonprofit institution that helps improve public policy through research and analysis. In 1995, RAND published a report on "The Potential of Nanotechnology for Molecular Manufacturing" by Max Nelson and Calvin Shipbaugh.

Private Foundations

- The Mayo Foundation/Mayo Clinic, Rochester, Minnesota

  Nanoelectronics research at the Mayo Foundation is focused in the Mayo Special Purpose Processor Development Group. This group is led by Dr. Barry Gilbert, who is one of the world's preeminent experts in the field of quantum-effect electronic devices nanoelectronics.

US Industry

- Hughes Electronics

  Hughes is engaged in nanoelectronics research. However, only limited information about Hughes' nanoelectronics activities is available on their corporate WWW pages.

- International Business Machines (IBM)

  IBM is one of the leaders world-wide in nanotechnology and nanoelectronics research. Some of their capabilities are on view in the • I B M Almaden Scanning Tunneling Microscope Image Gallery. This page contains spectacular images of atoms being manipulated with an STM.

Additional IBM nanotechnology research is performed at the IBM-Zurich Lab. One of the most important tools of nanotechnology, the scanning tunneling electron microscope (STM), the first • proximal probe device, was invented at IBM-Zurich. The Lab's Web site has a description of an IBM-Zurich molecular positioning experiment that uses an STM.

- Motorola, Inc.

  Motorola is among the leaders in nanoelectronics research. However, only limited information about Motorola's nanoelectronics activities is available on their corporate WWW pages.

- Texas Instruments, Inc. (TI)

The Texas Instruments Nanoelectronics Group is one of the world's leaders in nanoelectronics research. However, only limited information about TI's nanoelectronics research activities is available on their corporate WWW pages. Several of the TI nanoelectronics investigators are listed, however, on the Who's Who in Nanoelectronics page.

Japanese Laboratories

- ERATO Single Quantum Dot Project

  Five year, $20 million interdisciplinary project to investigate the basic properties and fabrication of quantum dots. Begun in October 1995. Sponsored by Japan's ERATO (Exploratory Research for Advanced Technology) Initiative, under the auspices of Japan's Science and Technology Agency.

- Fasol Lab at the University of Tokyo

  The Fasol Lab home page has an in-depth nanotechnology index.

- Hitachi Corp. Research and Development, Tokyo, Japan

  Hitachi is engaged in an extensive nanoelectronics research effort, involving a number of preeminent scientists and engineers. However, only limited information about Hitachi's nanoelectronics activities is available on their corporate WWW pages.

European Laboratories

- Cavendish Laboratory, Cambridge University, Great Britain

  As noted by Dr. William Tolles in his 1994 report "Nanoscience and Nanotechnology in Europe," important research on the subject of nanoscience and nanoengineering is ongoing at the Cavendish Laboratory. However, little information about Cavendish's nanoelectronics activities is available on their laboratory WWW pages.

- Project on Nanometre Structures for Future Optoelectronic Applications sponsored by the European Union's Espirit Program for Information Technologies Research
- Project on Organic Materials for Electronics (and Nanostructures) sponsored by the European Union's Espirit Program for Information Technologies Research
- Project on Resonant Tunneling for Novel Electronic Applications sponsored by the European Union's Espirit Program for Information Technologies Research
- University of Glasgow Nanoelectronics Research Center, Glasgow, Scotland
- IBM-Zurich Laboratory, Zurich, Switzerland

Much of the IBM's nanotechnology research is performed at the IBM-Zurich Lab. The Lab's Web site has a description of an IBM-Zurich molecular positioning experiment.

- National Physical Laboratory, Teddington, Great Britain

  As noted by Dr. William Tolles in his 1994 report "Nanoscience and Nanotechnology in Europe," important research on the subject of nanometrology is ongoing at the National Physical Laboratory. However, little information about the National Physical Laboratory's nanoelectronics activities is available on their laboratory WWW pages.

- University College London, UK

  Professors Terry Fountain and Michael Duff at the University College London are leading experts in the field of architectures for nanoelectronic computers. However, little information about the University College London's nanoelectronics activities is available on their university WWW pages.

Nanotechnology Research - Lab Groups

- Massachusetts Institute of Technoogy (MIT) - Microsystems Technology Laboratories - explores microsystems, microfabrication, and manufacturing technologies.
- Princeton University - NanoStructures Laboratory - explores and develops new nanotechnologies to fabricate structures; innovates nanoscale electronic, optoelectronic, and magnetic devices.
- New York University (NYU) - Seeman Lab - investigating unusual DNA molecules in model systems that use synthetic molecules, with a major effort in DNA nanotechnology.
- University of Cincinnati - Nanoelectronics Laboratory - research at or around the nanometer scale in semiconductors and other electronic materials.
- Academia Sinica - Surface Nanostructure Laboratory - creates, characterizes, and exploits nanostructures (1~50 nm) on solid surfaces.
- Waseda University - Ohdomari Lab - research on the process of nanostructure fabrication, and the physics of silicon surfaces and interfaces.
- University of Pune - Scanning Tunneling Microscopy Group - performs research on nanoparticles, nanolithography, and noise in STM.

NIH Nanotechnology and Nanoscience Information

This page contains information on (1) currently active NIH and BECON research and training opportunities and (2) listings of funded grants for NIH and BECON program announcements related to nanotechnology and nanoscience. Additional information on the NIH bioengineering funding opportunities can be found at http://www.becon.nih.gov/becon_funding.htm.

While much of biology is grounded in nanoscale phenomena, NIH has not re-classified most of its basic research portfolio as nanotechnology. Only those studies that use nanotechnology tools and concepts to study biology; that propose to engineer biological molecules toward functions very different from those they have in nature; or that manipulate biological systems by methods more precise than can be done by using molecular biological, synthetic chemical, or biochemical approaches that have been used for years in the biology research community, are classified as nanotechnology projects. NIH participates in the National Nanotechnology Initiative (www.nano.gov), the U.S. federal multi-agency R&D program for nanoscale science, engineering, and technology.

A list of technical and financial contacts for Nanoscience and Nanotechnology related projects is given at: http://www.becon.nih.gov/nano_contacts.htm.

A list of technical and financial contacts for PA-02-125, Bioengineering Nanotechnology Initiative (SBIR) is given at: http://www.becon.nih.gov/nano_contacts_sbir.htm.

Nanomedicine NIH Roadmap Initiatives:

- Nanomedicine Roadmap Homepage

NIH Reports Related to Nanotechnology:

- NIBIB/DOE Workshop on Biomedical Applications of Nanotechnology, March 17-18, 2005
- Nanoscience and Nanotechnology Symposium - June 25-26, 2000
- NHLBI Nanotechnology Working Group Report
- NCI Cancer Nanotechnology Brochure
- Recent Slideshow on NIH Nanotechnology Funding Opportunities/Examples

Other Trans-NIH Initiatives:

- Nanoscience and Nanotechnology in Biology and Medicine
- Bioengineering Nanotechnology Initiative
- Bioengineering Research Partnerships
- Exploratory/Developmental (R21) Bioengineering Research Grants

National Heart Lung and Blood Institute

- Innovative Technologies for Engineering Small Blood Vessels
- NHLBI Programs of Excellence in Nanotechnology

National Cancer Institute

- NCI Alliance for Nanotechnology in Cancer (including current funding opportunities)
- Innovative Technologies for Molecular Analysis of Cancer
- Application of Emerging Technologies for Cancer Research

National Institute of General Medical Sciences

- Single Molecule Biophysics and Nanoscience - Research on, and development of, new and improved instruments, methods, and technologies for nanoscience, and for the analysis of single protein and nucleic acid molecules and their complexes in vivo and in vitro. Current approaches include optical and fluorescent spectroscopies, scanning probe microscopy, and biomechanical techniques to analyze the behavior and heterogeneity of single molecules and subcellular structures at the nanometer scale. Examples of targets for study include protein or RNA folding, enzyme catalysis, signaling, molecular machines, and the assembly and dynamics of complex cellular structures. A major goal is to develop and enhance existing methods and reagents for the 3-D visualization of cellular processes in living cells in real time at high resolution.

National Toxicology Program (NTP)

Three agencies form the core of the NTP:

- National Institute of Environmental Health Sciences of the National Institutes of Health (NIEHS/NIH)
- National Institute for Occupational Safety and Health of the Centers for Disease Control and Prevention (NIOSH/CDC)
- National Center for Toxicological Research of the Food and Drug Administration (NCTR/FDA)
- The NTP is conducting studies on the toxicology of nanoscale materials.
- Report of the first nanotoxicology workshop http://www.nanotoxicology.ufl.edu/ aimed at developing appropriate experimental strategies for evaluating the safety/toxicity of nanomaterials

Additional NIH Nanoscience and Nanotechnology Research Funding Opportunities:

- DEVELOPMENT OF CELL-SELECTIVE TOOLS FOR STUDIES OF THE BLADDER, PROSTATE, AND GENITOURINARY TRACT - Application Receipt Dates: February 1, 2005

- NOVEL APPROACHES TO CORNEAL TISSUE ENGINEERING - Application Receipt Dates: February 1, June 1, October 1, ending June 1, 2005
- NINDS EXPLORATORY/DEVELOPMENTAL PROJECTS IN TRANSLATIONAL RESEARCH - Application Receipt Dates: February 1, June 1, October 1, ending July 31, 2005
- SMALL BUSINESS BIODEFENSE PROGRAM - Application Receipt Dates: April 1, August 1, December 1, ending Aug. 2, 2005
- CUTTING-EDGE BASIC RESEARCH AWARDS (CEBRA) - Application Receipt Dates: February1, June 1, October 1, ending Oct. 5, 2005
- NEUROTECHNOLOGY RESEARCH, DEVELOPMENT, AND ENHANCEMENT - Application Receipt Dates: February 1, June 1, October 1 ending October 2006

Information on Funded Grants and Projects for NIH/BECON Program Announcements:

- Summary of Funded NIH Bioengineering Nanotechnology Initiative (SBIR)
- Summary of Funded NIH Bioengineering Nanotechnology Initiative (SBIR)
- Summary of Funded NIH Bioengineering Nanotechnology Initiative (SBIR)
- Summary of Funded NIH Bioengineering Nanotechnology Initiative (SBIR)
- Summary of Funded NIH Bioengineering Nanotechnology Initiative (SBIR)
- Summary of Funded Nanoscience and Nanotechnology in Biology and Medicine (PAR-03-045)

NSF - NSES

The potential impact of research on the nanoscale is well recognized, and yet to remain nationally and internationally competitive formidable challenges must be met. Through a unique partnership of academic institutions, industrial partners, and national labs, the NSF Nanoscale Science and Engineering Center (NSEC) for Integrated Nanopatterning and Detection Technologies, headquartered at Northwestern University under the leadership of Chad A. Mirkin, is helping to meet these challenges by providing a cross-disciplinary research program well integrated with innovative educational outreach programs.

The NU-NSEC brings together recognized leaders from Northwestern University, University of Chicago, University of Illinois/Urbana-Champaign, and Argonne National Laboratory, who, building on the nanopatterning tools and synthetic methods unique to this Center, are making advances in the development of advanced nanopatterning techniques and nanoscale sensors. Potential applications in disease and biological/

chemical detection systems arena are expected to be profound. However, this research will also directly impact new technological directions outside of the existing purview of the Center including molecular electronics, catalysis, information storage, and therapeutics.

In a field as dynamic as nanotechnology, collaborations are a vital means to transfer knowledge and an important component of the Center mission. NSEC researchers have numerous meaningful collaborations with industry, national laboratories, and academia. Additionally, the Nanotechnology Corporate Partners (NCP) program provides a forum for companies and business groups with allied interests to develop partnerships with the NSEC and help transition its technology into the private sector. Working with the Kellogg School of Management, the NSEC also offers the Small Business Evaluation and Entrepreneurs (SBEE) program, which gives scientists and engineers with new technologies assistance with development of comprehensive business plans for presentation to potential investors. A $10 million seed fund was established by Lurie Investment Fund to initiate business development based on technology developed at the Center, and three companies have already been successfully launched.

Excellent in-house user facilities are available to NSEC researchers. Additionally, the State of Illinois committed $5,000,000 to support the development of next-generation instrumentation and to establish a multi-user, multi-purpose, open-access Nanoscale Imaging, Fabrication, Testing and Instrumentation (NIFTI) facility to provide soft lithographic writing, imaging, and analysis capabilities.

The NU-NSEC is also committed to educating the scientists, technicians, and teachers of tomorrow. This commitment is realized not only in the research of graduate students, undergraduates, and postdoctoral associates, but also through the Center's innovative *educational programs.* Some of these programs build on current educational strengths and infrastructure, while others launch new initiatives both internally and through exciting new partnerships with the Museum of Science and Industry, Chicago, and Harold Washington College (a primarily minority institution).

Undergraduates and minority undergraduates from around the country have opportunities to engage in research at the Center. High school science teachers, many from schools with large minority populations, participate in an innovative program which combines hands-on research with curriculum development. NSEC researchers are collaborating with pre-college educators on the development of a "Nanoscience Module" to introduce nanoscience concepts at the high school level through guided hands-on activities and design challenges. The NSEC is also collaborating with the Museum of Science & Industry, Chicago, to produce a design concept for a 13,000 square foot exhibit in nanotechnology, which is expected to open in 2006 and educate millions of museum visitors each year. An interactive web site for children entitled "Mission Spec-tacular" will be launched in spring 2003, and a series of dramatic educational demonstrations entitled "Frontiers of Nanotechnology" has already been premiered with pre-college and lay audiences. NSEC researchers and their students are also actively involved and highly visible in the broader community, through lectures, radio, and television.

Through synergistic research of the highest quality, strong and active collaborations, and innovative outreach programs, the NU-NSEC is advancing scientific knowledge and understanding in the area of integrated nanopatterning and nanoscale detection technologies, while advancing formal education as well as public understanding of nanoscale science and engineering.

Nanoscience and Technology at US - DOE Labs

The Role of DOE in the National Nanotechnology Initiative:

"The DOE has a stunning portfolio of research and scientific user facilities devoted to visualizing, characterizing, and controlling the nanoworld - from atoms and molecules to bulk materials - which makes the Department's research capabilities unique in the world. The DOE is currently making a broad range of contributions in these areas. For example, the enhanced properties of nanocrystals for novel catalysts, tailored light emission and propagation, nanocomposites and supercapacitors are all being explored. Nanocrystals and layered structures offer unique opportunities for tailoring the optical, magnetic, electronic, mechanical and chemical properties of materials, and DOE researchers are have synthesized layered structures for electronics, novel magnets, and surfaces with tailored hardness.

Specific examples of past accomplishments at DOE include:

- Addition of aluminum oxide nanoparticles that converts aluminum metal into a material with wear resistance equal to that of the best bearing steel
- Novel optical properties of semiconducting nanocrystals that are used to label and track molecular processes in living cells
- Nanoscale layered materials that can yield a four-fold increase in the performance of permanent magnets
- Layered quantum well structures to produce highly efficient, low-power light sources and photovoltaic cells
- Novel chemical properties of nanocrystals that show promise as photocatalysts to speed the breakdown of toxic wastes
- Meso-porous inorganic hosts with self-assembled organic monolayers that are used to trap and remove heavy metals from the environment

The DOE also maintains a large array of major national user facilities that are ideally suited to nanoscience discovery and to developing a fundamental understanding of nanoscale processes. Large computational facilities at DOE will also be key contributors in nanoscience discovery, modeling and understanding.

National Nanotechnology Initiative Research at the DOE:

Major new efforts in nanoscale science, engineering, and technology at the Department of Energy will take advantage of opportunities afforded by recent advances. These efforts will be part of the Basic Energy Sciences (BES) program and have the following broad goals:

1. to attain a fundamental scientific understanding of nanoscale phenomena, particularly collective phenomena;
2. to achieve the ability to design and synthesize materials at the atomic level to produce materials with desired properties and functions;
3. to attain a fundamental understanding of the processes by which living organisms create materials and functional complexes to serve as a guide and a benchmark by which to measure our progress in synthetic design and synthesis; and
4. to develop experimental characterization tools and theory/modeling/simulation tools necessary to drive the nanoscale revolution.

The principal missions of DOE in science, energy, defense, and environment will benefit greatly from developments in these areas. For example, nanoscale synthesis and assembly methods will result in significant improvements in solar energy conversion; more energy-efficient lighting; stronger, lighter materials that will improve efficiency in transportation; greatly improved chemical and biological sensing; use of low-energy chemical pathways to break down toxic substances for environmental remediation and restoration; and better sensors and controls to increase efficiency in manufacturing."

Sandia National Labs

It's no wonder that venturing into the tiny domain of atoms and molecules is catching on. Amazing discoveries are being made, inspiring scientists the world over.

Throughout the scientific community, including Sandia National Laboratories, researchers say building things atom-by-atom or molecule-by-molecule will revolutionize the production of virtually every human-made object. *Exciting* prospects—but they also point out the promise of nanotechnology can *only* be realized if we learn to understand the special rules that control behavior at this small scale and develop the skill needed to integrate these concepts into practical devices.

The excitement stems from the understanding that the behavior of materials at the nanoscale is *nothing* like that at the large scale. Only recently have the necessary tools, such as powerful new microscopes, been developed to let researchers see these surprising behaviors. Sandia National Labs' scientists and engineers are among the leading architects and builders of these tools. To appreciate their unique design features, it helps to get an idea of the size of things at the atomic level. In nanoscience objects are measured in nanometers, 1 billionth of a meter. For comparison, the smallest features on current computer chips, measure about 200 nanometers. And a human hair is 100,000 nanometers thick.

Moving beyond observation, scientists are now poised to make exciting advances in nanotechnology, the creation of materials, devices and systems through the *control* of matter at the atomic level. By understanding and controlling the way molecules organize into nanoscale patterns, scientists are discovering new phenomena and learning to design materials with vastly different sets of properties. As one Sandia scientist put it "Design possibilities are limited only by one's imagination."

Sandia Labs continues to enhance its proficiency in many fields, all in keeping with its Department of Energy mission to unite science and engineering to serve national needs. And along with world class capabilities in materials science, micro fabrication - including 40,000 square feet of clean room space - high performance computing and systems engineering - Sandia is uniquely positioned to be the integrating center for new discoveries in nanoscience.

Sandia is uniquely positioned to be the intergrating center for new discoveries in nanoscience. Nano materials are already helping to attain scientist's vision for new technology such as, the *micro-chemlab*, by increasing a thousandfold the ability to collect the chemicals of interest. Currently in the development stage, this handheld device will carry out all of the functions of a full chemistry laboratory allowing first responders, military, and emergency personnel to rapidly access chemical and biological hazards.

All of these advances are coming from the work of multi-disciplanary teams. Their ability to bring together and integrate micro-scale devices is another key part of the nanotechnology success story Through strong commitment and continued partnerships with other national labs, universities and industry. Sandia scientists believe the promise of nanotechnology applications that touch peoples lives can be realized. Grand prospects based on tiny environments. It will require thinking big in a nano sized world.!

U.S. Naval Research Lab: Nanotechnology Science and Technology

Strategic Research Area

- To achieve dramatic, innovative enhancements in the properties and performance of structures, materials, and devices that have controllable features on the nanometer scale (i.e., tens of angstroms).

- The ability to affordably fabricate structures at the nanometer scale will enable new approaches and processes for manufacturing novel, more reliable, lower cost, higher performance and more flexible electronic, magnetic, optical, and mechanical devices.

Recognized Applications of Nanoscience

- Ultrasmall, highly parallel, computers with multi-teraflop speed
- Image information processors, e.g., extraction and recognition
- Low-power personal and autonomous communication and computation devices
- high-density information storage devices, e.g., terabit/cm2 nonvolatile memory
- Lasers and detectors for weapons and countermeasures
- Optical (infrared, visible, ultraviolet) sensors for improved surveillance an targeting
- Integrated sensor suites for chemical and biological agent detection
- Catalysts for enhancing and controlling energetic reactions
- Synthesis of new compounds (e.g., narrow-bandgap materials)
- Designer materials with combinations of properties that do not currently exist

SRA Representatives

Air Force: Dr. Gernot S. Pomrenke

Army: Dr. Henry Everitt

DARPA: Dr. Christie Marrian

Navy: Dr. James Murday (Chair)

US Office of Naval Research

The Surface Nanoscience and Sensor Technology section of the Naval Research Laboratory specializes in (1) basic and applied research in surface science, the mechanical properties of materials, the molecular basis of adhesion, film nucleation and growth, and chemical/biological sensors; (2) development of tip-based probes for studies in the area of nanostructures and nanomechanics of surfaces and single molecules; and (3) development of new surface/interface analysis techniques for structural and chemical analysis of materials.

Current Areas of Research

Structure of semiconductor surfaces and interfaces

We are studying the atomic-scale structure and reactivity of single crystal silicon, germanium, and III-V compound semiconductor surfaces and interfaces in ultra-high vacuum (UHV) using scanning tunneling microscopy (STM) and other surface analysis techniques. Due to the rapidly shrinking size of electronic devices, such studies are vital to the development of future Navy electronics.

Nanomechanics of surfaces

The nanomechanical properties of materials has become an area of increasing importance as the scale of coatings, multilayer films, particles, components and devices approach atomic dimensions. A goal of our nanomechanics program is to replace the phenomenological understanding of adhesion, tribology and interfacial/fracture mechanics with specific atomic- or molecular-scale mechanisms. Our research objectives include measuring the intermolecular and surface forces of materials with molecular specificity and resolution; elucidating the mechanics of materials or species under compression, tension, or shear; and developing new science-based molecular models for adhesion and tribology. This program also focuses on the relationship between interfacial chemistry and the structural and mechanical properties of interfaces by applying state-of-the-art nanoprobe technology including atomic force microscopy (AFM) and nanoindenation techniques.

Interaction forces of biomolecules

Atomic force microscopy (AFM) offers the unique ability to perform nanomechanical measurements on individual biological molecules. We have developed this ability and are using it to investigate inter- and intra-molecular interaction forces in model systems, between complementary strands of DNA, and between antibodies and their antigens. Molecular modeling studies are carried out to provide insights into the bond rupture mechanisms that underly these experiments.

FABS / BARC / FDA biosensors

Building on our research into the interaction forces between biomolecules, we are developing sensors for biomolecules and other chemicals, viruses, and bacteria. These sensors use microfabricated transducers to detect the minute forces with which antibodies bind to particular molecules. This research will result in the development of portable, highly-sensitive, automated chemical and biological sensors. The Force Differentiation Assay (FDA) and the Force Amplified Biological Sensor (FABS) are expected to deliver orders-of-magnitude sensitivity improvements over methods that are now in common use. The Bead Array Counter (BARC) will operate on similar principles but will trade sensitivity for the ability to simultaneously detect hundreds of types of molecules.

Quantitative surface analysis

The surface analytical techniques of x-ray photoelectron spectroscopy (XPS) and Auger electron spectroscopy (AES) are used to quantitatively determine the surface composition or in-depth distribution of low-level dopants in a variety of advanced materials including semiconductors, polymers, diamond films, and engineering alloys. These approaches are employed also for tribology, surface cleaning, and insulation degradation studies. Experimental and theoretical methods have been developed to overcome problems caused by matrix effects, surface segregation, background, peak overlap, specimen charging, sputtering artifacts, etc. Auger lineshapes and factor analysis methods are used extensively to extract chemical information.

Langmuir-Blodgett and self-assembled films

Research is conducted on physicochemical properties and effects of surface-active organic compounds that 1) form organized monomolecular films and 2) can be transferred to solid surfaces as mono- or multilayer films with precisely controlled properties such as packing density and phase state. Such films are also used in fundamental research studies as reference standards for scanning probe microscopes; as model systems to help understand the physical and chemical properties of biological membranes; and as model systems to increase understanding of the molecular basis of adhesion and lubrication. They have also been used in practical applications that take advantage of the unique properties of organized thin films, for example, as rapidly-equilibrating active coatings for gas sensors. Properties and effects of films that occur naturally on the sea surface have also been studied.

Facilities

Scanning probe microscopes

Scanning tunneling microscopy/spectroscopy (STM/S) enables the surface topography, chemical reactivity, and electronic structure of conductive substrates to be observed with atomic-scale resolution. Atomic force microscopy (AFM) can provide nanometer-scale resolution of surface topography, mechanical properties, and tip-surface interaction forces on both conductive and insulating substrates. The tip-surface interaction forces, including frictional forces, can be measured with nanonewton (single chemical bond) resolution.

- Homebuilt ultra-high-vacuum (UHV) STM/S with facilities for low-energy electron diffraction (LEED) and Auger electron spectroscopy (AES), for characterizing atomic-scale surface properties;
- Homebuilt STM/S with fast, variable-temperature STM and facilities for LEED, AES, temperature programmed desorption, for real-time atomic-scale observations of surface reactions;

- Park Scientific Instruments AutoProbe UHV STM/AFM integrated with the NRL Molecular Beam Epitaxy (MBE) Epicenter for characterizing semiconductor surfaces following MBE and related processing;
- RHK Technology STM 2000 UHV STM/S;
- Homebuilt ambient STM/S with capability of generating high-frequency surface acoustic waves (SAW) for measurement of elastic properties of surfaces;
- Nanoscope IIIa multimode AFM (lateral force, magnetic force, and tapping modes) equipped with breakout box and Hysitron TriboScope nanoindentation instrument for mechanical property measurements;
- Nanoscope IIIa multimode AFM (lateral force, magnetic force, and tapping modes) equipped with breakout box and force-volume mapping system;
- Homebuilt ambient AFM for adhesion and tribology studies;
- Homebuilt liquid-cell AFM for molecular recognition studies;
- Topometrix TMX 2000 AFM, belonging to code 6174, but housed within code 6177.

Surface analytical techniques

These techniques are used to characterize the surfaces of solids with respect to their electronic and geometric structure, mass, molecular weight, or optical properties. The techniques include X-ray photoelectron spectroscopy (XPS) and Auger electron spectroscopy (AES). Both approaches carry information about the electronic structure of the solid and its chemical composition. In addition, a rastered electron beam (AES) or parallel imaging (XPS) can produce images of chemical species.

- Perkin-Elmer 660 50 nm scanning Auger microprobe (SAM) with sputter depth profiling capability and X-ray dispersive analyzer, for analyzing thin films and determining the lateral or in-depth distribution by depth profiling of elements at the <0.5 atomic percent level.
- Surface Science Instruments ESCA301 150 μm monochromatic XPS with depth profiling capability and angle-resolved analysis.
- VG Scientific model 220i-XL XPS with monochromatic and non-monochromatic X-ray sources, depth profiling capability, small area analysis (20 μm) and <5 μm spatial resolution parallel image capability.
- JA Woollan M44 410-750 nm ex situ spectroscopic ellipsometer with ambient and liquid cells for film thickness, surface roughness, optical constant, and composition measurements.

Chemistry laboratories

These facilities are primarily intended for the preparation and characterization of 1) antibody or DNA-derivatized surfaces and 2) Langmuir-Blodgett and self-assembled films.

- Biochemistry and molecular biology facilities with -70° C freezer, UV/vis spectrophotometer, PCR cycler, gel box, microplate reader, incubator, glove box, centrifuge, etc.
- Zeiss Axiovert and Axiotech reflected- and transmitted-light optical microscopes with epifluorescence and differential interference contrast (DIC) attachments; equipped with Narashige three-dimensional hydraulic micromanipulator and video microscopy/digital video capture system.
- Dedicated surface chemistry clean-room laboratory with KSV 5000 Langmuir-Blodgett System.

Magnetics equipment

The following equipment for generating and measuring magnetic fields is part of the FABS/BARC/FDA program.

- Group 3 DTM-141 Digital Teslameter with serial communications and MPT-141 miniature Hall probe.
- Walker Scientific 40 kOe pulsed magnetizing coil with 1 inch bore.

Scanning electron microscopy, Fourier transform IR and Raman spectroscopy, solid state NMR, X-ray fluorescence, Rutherford backscattering, MS/MS, and other supporting techniques are also available within the Chemistry Division.

W.M. Keck Nanostructures Laboratory - University of Massachesetts, Ahmerst

The W.M. Keck Nanostructures Laboratory is a part of the Materials Research Science and Engineering Center (MRSEC) and the Silvio O. Conte National Center for Polymer Research at the University of Massachusetts at Amherst. It is equipped with several Atomic Force Microscopes, Optical Microscopes, Variable Angle Spectroscopic Ellipsometer (VASE), Thin Film Optical Profilometer, Stylus Profilometer, Metal Evaporator etc. for the use of UMass community and as well as from other academic institutions and Industry. The mission of the Keck Nanostructures Laboratory is to provide access to material characterization equipment, technical support, training and consultation, as well as to perform a range of services for users in the area of Atomic Force Microscopy (AFM), Small Angle X-ray Scattering (SAXS), Variable Angle Spectrocopic Elliposmetry (VASE) and Optical Microscopy. The laboratory capabilities include high-resolution imaging of materials structure, analysis of X-ray Diffraction patterns, Thin Film characterization etc. In addition the Nanostructures Laboratory has access to the Confocal Laser Scanning Microscope from the Optical Microscope facility of the MRSEC

NanoTech User Facility at the UW (NTUF)

Training

Welcome to the NanoTech User Facility (NTUF). NTUF is available to graduate students and industrial researchers as a cost center. NTUF staffs will provide training on instruments upon request. Please register online and we will contact you to schedule training sessions.

Training consists of two sessions. First, we will provide an in-depth introduction on the principles of the instrument and demonstrate capabilities by running standard samples. The second training session will be hands-on and users are encouraged to work with their own samples. We may request additional practice sessions depending on qualification on the use of equipment.

For qualified users who can perform independent experiments and who have agreed with NTUF policy, NTUF is available 24 hours a day and seven days a week.

- Zeiss Confocal Microscope Guided Tour
- Principles of Confocal LSM
- Beam Path Configuration Guide
- Zeiss LSM 510 Operating Procedure
- Volocity Users Guide
- Digital Multimode III Operating Procedure
- Nanoscope III User Manual (© Digital Instruments)
- Nanoscope III schematic (© Digital Instruments)
- Leica Operating Procedure
- Quick Procedure for Raman Spectroscopy
- Raman system schematic
- FEI Sirion Operating Procedure
- E-beam Lithography Training Procedure I
- E-beam Lithography Training Procedure II
- E-beam Lithography Introduction
- E-beam Resist (PMMA) Data Sheet (© MicroChem)
- NTUF E-Beam Resist Information

- NTUF Master Template I
- NTUF Master Template II

Charles B. Musgrave

Quantum Simulations of Molecular Processes in Chemical Engineering

Our research program focuses on using quantum mechanics to simulate molecular processes of important engineering problems. Our approach is fundamental and interdisciplinary and combines quantum mechanics, chemical kinetics, surface chemistry and materials chemistry. Our goals are to develop a fundamental molecular and mechanistic understanding of the processes underlying important new technologies and to establish quantum chemical simulations as a practical engineering tool for computational prototyping of systems with molecular detail. Systems we investigate include atomic layer deposition, dye sensitized nanostructured solar cells, solid oxide fuel cells, hybrid interfaces between semiconductors and organic and biological molecules, molecule-surface electrical junctions and nanotechnology.

Atomic Layer Deposition

ALD is a process that has recently gained enormous attention. It involves exposure of surfaces to alternating pulses of different precursors to form atomic layers of deposited material through self-limiting surface reactions. Currently, the most active area of ALD research is the deposition of high-K gate dielectrics, such as ZrO 2 and HfO 2 because finding a replacement for SiO 2 gate dielectrics is the most critical technical challenge facing the semiconductor industry in the current transition to nanoelectronic devices. Although research on ALD of high-K materials has recently exploded, no detailed mechanisms or first principles studies had been reported until we reported detailed mechanisms for ALD of several important materials including ZrO 2, HfO 2,, HfN, Al 2 O 3, SiO 2 and Si 3 N 4. In addition to being the first group to describe detailed mechanisms of ALD, we are creating the fundamental framework for the understanding of the ALD process which provides a based upon which ALD can be extended and improved. We are also exploring reactions for low-temperature ALD on organics and self-assembled monolayers for area-selective ALD and molecular electronics, ALD for fuel cells, and dye sensitized solar cells.

Dye-Sensitized Solar Cells

The heart of dye-sensitized solar cells is the dye-molecule bound to a metal oxide nanostructured surface who's primary function is to covert photons to electron hole pairs in order to harvest solar energy. The goal of our work on dye-sensitized solar cells (DSC) is to develop new molecular technologies to control the key interfacial properties of DSC with the aim of improving their efficiency, and reducing their cost and the energy required for their manufacture. Our strategy is to use quantum chemical simulations and carrier transport simulations to perform molecular engineering of the

DSC interface structures by computationally prototyping both the interfacial physics which governs the operation of the DSC as well as the chemical processes involved in their fabrication. The DSC has great potential for economically harvesting significant amounts of the solar energy incident upon the earth's surface without the production of green house gases. Furthermore, despite being early in their development cycle relative to traditional photovoltaics, much progress has been made in both the fabrication and efficiency of the DSC with efficiencies exceeding 10%.

Our view of DSC is that the surface-tethered dye is an optoelectronic molecular device whose function critically depends on the quantum electronic structure. From our previous work in organic functionalization of semiconductors surfaces and ALD of metal oxides we have become one of the world's leading groups in the electronic structure of organic functionalized surfaces as well as in the electronic structure of organometallic molecules on metal oxide surfaces and we leverage this expertise to enable us to expand the theory and understanding of the DSC with the hope that this will lead to improvements and subsequent faster implementation of DSC technology.

Low Temperature Solid Oxide Fuel Cells

Fuel cells have the potential to play an immensely important role in a potential hydrogen economy. Solid oxide fuel cells (SOFC) use solid electrolyte gates and have many advantages over polymer gate systems. The main disadvantage is that they must be operated at very high temperatures (1000-1200 C). Our goal is to simulate oxygen ion migration through the electrolyte to determine the migration mechanism and to design and computationally prototype new electrolytes with higher ion conductivities with the goal of reducing the operating temperature of the fuel cell by using dislocations, and built-in chemical potential gradients, and ultra thin membranes fabricated using ALD.

Hybrid Bio and Organic Semiconductor Devices

Semiconductors are at the center of many key technologies including computation, communication, data storage and sensing. Inorganic interfaces in semiconductor based devices have been key components in the unique properties that these devices exploit for their operation. We have undertaken an extensive study of how various organic molecules react on semiconductor surfaces and developed the most comprehensive theoretical description of organic reactions on semiconductors. We are extending this leadership position in two ways. First, by moving from studying the reactivity of organics on semiconductors to investigating the unique electronic properties of the interfaces formed between various organic functional groups and semiconductor surfaces. Second, by extending the types of organic molecules on semiconductors we study to biological molecules, such as amino acids. We have recently found several interesting new types of reactions between biological molecules and semiconductor surfaces which are not only unique in their reactivity, but which also exhibit strong quantum mechanical coupling between the biological molecule and the semiconductor, potentially leading to new types of electrical junctions in molecular electronics. We are also exploring the use of this phenomenon to electrically connect carbon nanotubes to each other.

Molecular Electronics

There has been significant recent interest in using molecules as components in electronic circuits. Critical to this area is the ability to transport electrons between molecules and bulk materials across the interface formed between the molecule and the material surface. This is mostly governed by the electronic structure of the molecule-material interface. Several of our projects involve charge transfer between molecules and surfaces including dye-sensitized solar cells, the oxidation and reduction in fuel cells and our quantum resonance junctions between certain amino acids and Si and Ge surfaces. Our strategy is to use quantum simulations to develop a clear understanding of the charge transfer process so that we may engineer molecules to optimize them as electronic junctions.

Theoretical Studies of a Hydrogen Abstraction Tool for Nanotechnology

Processes that use mechanical positioning of reactive species to control chemical reactions by either providing activation energy or selecting between alternative pathways will allow us to construct a wide range of complex molecular structures. An example of such a process is the abstraction of hydrogen from diamond surfaces by a radical species attached to a mechanical positioning device for synthesis of atomically precise diamond-like structures. In the design of a nanoscale, site-specific hydrogen abstraction tool, we suggest the use of an alkynyl radical tip. Using *ab initio* quantum-chemistry techniques including electron correlation we model the abstraction of hydrogen from dihydrogen, methane, acetylene, benzene and isobutane by the acetylene radical. Of these systems, isobutane serves as a good model of the diamond (111) surface. By conservative estimates, the abstraction barrier is small (less than 7.7 kcal mol^{-1}) in all cases except for acetylene and zero in the case of isobutane. Thermal vibrations at room temperature should be sufficient to supply the small activation energy. Several methods of creating the radical in a controlled vacuum setting should be feasible. Thermal, mechanical, optical and chemical energy sources could all be used either to activate a precursor, which could be used once and thrown away, or alternatively to remove the hydrogen from the tip, thus refreshing the abstraction tool for a second use. We show how nanofabrication processes can be accurately and inexpensively designed in a computational framework.

Tim Harper

Interview with Tim Harper

Founder & CEO of the nanotechnology integrator CMP Cientifica, Tim is a former European Space Agency engineer, who now runs conferences and information services and works closely with the VC industry to help select and guide investments in nanotechnology companies. Tim is co-author of the *Nanotechnology Opportunity Report* and joint editor in chief of *The X Report*. A brief biography of Tim Harper is available on the CMP website.

Steve: What's the story behind CMP-Cientifica, Trends in NanoTechnology (TNT) and your new collaboration agreement with Spark Inversiones?

Tim: What we are trying to achieve with CMP, TNT and Spark are three separate activities, although they are all involved in nanotechnology. In essence they are the technologies, the information and the funding. I should make it clear that CMP Cientifica and Spark Inversiones are two very separate companies that are collaborating in the field of nanotechnology, while the TNT label covers all of CMP Cientifica's information services, whether conferences or publications.

CMP Cientifica has been involved in nanofabrication for a long time, since I was at ESA in fact. Over the last four years we have worked with both clients & suppliers to develop tools and techniques with nanometer precision. Unfortunately, most of our work is covered by NDA's so I can't tell you as much as I would like to.

CMP Cientifica also coordinates the Phantoms network, an EU funded network on Nanoelectronics and NanoSpain, a more local network. Our experience shows us that the best way to advance technology is to bring together a wide range of experts, and get them to talk to each other. That's where the problems get solved, and it helps reduce duplication of effort. The TNT conferences are a great enabler for cooperation, and we have an excellent relationship with the NSF in the US, that allows us to get a truly global perspective on nanotechnology.

TNT Weekly was born out of the realization that most people outside the scientific community had almost no idea what nanotechnology is. In most people's minds, nanotechnology is tangled up with miniature robots, grey goo and terraforming Mars. You may think that the science fiction associations don't matter, but if you are looking for funding for a nanotechnology related business, it means you'll have a tough time convincing anyone to invest. VCs aren't in the business of investing in 30-year R&D projects, they need shorter term returns on their investments, but that's not the way most VCs see nanotechnology.

Essentially what we are doing with TNT Weekly is trying to provide clear, unbiased factual information about nanotechnology. The fact that it is as popular with the investment community as with the scientific community indicates we're doing something right.

Given that we have the technologies, and the information, the only remaining obstacle is getting funded. Most VC companies are staffed by financial professionals, and they can bring in analysts for whatever market sectors interest them. With nanotechnology there is no market to speak of, so how do you get a handle on where to invest? In fact you don't, and lack of information raises the risks, so the investment goes into better-understood markets.

What we are doing with Spark is bringing together our knowledge and experience of nanotechnology, with their expertise in early stage investing. This means that you'll be presenting your business plan to people who understand your market, and that's crucial for any kind of business success. Imagine that, a VC who really understands nanotechnology - for me that's a dream come true, and we hope it will be for European nano entrepreneurs.

Steve: What sort of environment, discussion and progress was there at TNT 2000 and what can be expected from TNT 2001?

Tim: The TNT conferences can be divided into two themes, science & networking. We keep these running together by having reasonably short lectures, followed by long coffee breaks & lunches. Organizing the conferences in old monasteries in medieval walled cities such as Toledo and Segovia certainly helps the atmosphere. What I always like to see is little knots of people frantically discussing fine points of technology, if you can get this going then the conference will be a success. TNT 2000 was our first conference on nanotechnology, and the proceedings have just been published in the IOP journal "Nanotechnology." This year we have a much more global participation, with support from both the EU and the NSF, and we've noticed a lot more industrial participation, which is vital if we are to get nanotechnology out of the lab and onto the market. TNT 2001 will also feature a lot more breakout meetings during the coffee breaks, where people can discuss specific themes.

What is a Nano?

Steve: If at all possible, how can one distinguish between a nanotechnology company and an ordinary company that is only using nano as a marketing scheme? I mean, anyone can say, "Our products are atomically precise."

Tim: Well the 'N' word seems to be in fashion at the moment, and we're not afraid to highlight misuses in TNT Weekly. Some of the marketing guys have already realized that a 'pico' is smaller than a 'nano', and it won't be long before we hear more about femto. Erik Moderegger of AT&S compiled a list of nano trademarks, which we put on our web site, and some of them are mind-boggling. I'd bet that very few of them are registered to anyone connected with nanotechnology.

As nanotechnology will have such an impact on most industries, the only way to distinguish nano from non-nano is by defining nanotechnology so tightly that it excludes everyone except Zyvex. Maybe I should turn the question around and ask whether a cosmetics company manufacturing nanoparticles for use in its products is a nanotechnology company or not?

Steve: In deciding which companies and products to include in my collection of links and how to organize them, I must draw the line somewhere. I consider nanotechnology to require observing, understanding and controlling the structure of matter on the nanoscale in order to get novel effects from the materials and subsequent integration of those materials. For instance, I would consider Nanophase to be a nanotechnology company because their sun-screens incorporate particles of a strategic size such that they absorb UV rays, without absorbing the larger visible frequencies. Thus, they block the harmful rays while remaining invisible to the human eye.

They don't carve each of their nanoparticles with nanomachines, but rather produce them with conventional methods. However, we now have novel analytical tools for observing nanoscale features, allowing us to tune the fabrication process to create the nanostructures that produce desired effects. Together with the theory, it then becomes possible to design devices based upon nanoscale or quantum size effects. Thus, we seem to be able to create nanostructured products based on rational design rather than years of trial and error. A couple more examples of real products resulting from this trend are molecular imprinting, solid lubricants and nearly perfect mirrors, as well as the instruments that allow nanoscale quality control of these objects.

I would not consider Zyvex a nanotechnology company, but rather a "Molecular NanoTechnology (MNT)" company, as they also call themselves. Perhaps MNT is a subset of the more general nanotechnology that you and I seem interested in, but it definitely deserves a class of its own.

Tim: Agreed. This just illustrates the difficulty of separating nano from non-nano. Even the definitions used by the EU and the NSF are slightly different; so, although we have to draw a line somewhere, its position has not yet been agreed on.

Revolutionary Nanotech Products

Steve: I imagine that with nanotechnology it might be possible to create a liquid paint that once it dries could function as a display screen. The paint might either be subsequently plugged in to receive signals and energy, or simply be solar powered and receive wireless signals. Of course, it needn't emit light as our current display screens do. It would be enough just to have pixels that could change color. Might nanotechnology economically produce such a paint?

Tim: Nanotechnology will be bound by the same rules of economics as most other companies, selling various things to other people for a profit. Our experience in the Field Emission Display market has shown the difficulties of taking a great idea to market, ten years of development have yet to yield a product that can be sold at a reasonable price. As an investor, I would be very wary of anyone proposing a paint that functions as a screen, is self powered, communicates with other screens, tap dances and a whole host of other things. For each of those applications, you have a different set of problems, of lifetime, of stability, of interfacing, of driving the thing & so on.

If someone comes along with a plan to produce a display based on nanoparticles, that's fine. When they start adding all kinds of other functions before the core product has been developed, it becomes a problem. This comes down to the age-old fight between engineers and business people. The reason for this is driven by economics rather than science and engineering, and the need to have, at some clearly defined time-scale, a marketable product. Engineers and scientists will always give you something better in a few months time, that's their job. However, companies, and their investors, need to take a decision on how good something needs to be before you can put it on the market. As Guy Kawasaki says, 'Don't Worry, Be Crappy" - meaning get a product on the market in order to generate some sales income to pay for the fixes/improvements. You could argue that Microsoft takes this to extremes by getting an unfinished product onto the market as quickly as possible and then adding the patches at their leisure. If I were to be cynical, I could mention the premium rate lines for technical support in this context.

So the answer is, nanotechnology might produce such a paint, but you would probably have to commercialize the applications one step at a time. There is nothing wrong with looking further ahead in terms of functionality and applications, but at some stage, and preferably an early stage, you have to have some kind of product that you can move from development to sales.

Steve: Perhaps my imagination was running a little too far into the future. The first step towards smart paint would clearly just be a paint that could change color on command. Of course, such materials exist, but there isn't really a market for them. Thus, there is nothing motivating the traditional step-wise development of smart paint - unless, the whole product could be strategically designed in advance, based upon existing data stored in steadily shrinking computer chips... How long do you expect until we begin to see some revolutionary nanotech products?

Tim: Reading the NSF report on the societal implications of nanotechnology, it amazed me that so much of the thinking was linear, and based on our current understanding of technology. While nanotechnology may make some things better it will also create whole classes of new materials, and consequently markets, that we would be crazy to try to predict right now. Because of this wide spectrum of applications, there are some nanotechnology products that have been commercialized for years now, such as nanometer sized gold particles for labeling of oligonucleotides, lipids, peptides, proteins etc., and other applications still bubbling away in University labs that might bear fruit in 20 years. So that's our time-scale, any time from now, until sometime in the future!

Steve: It seems that the demand for smart paint (I want smart paint!) is relatively unimportant compared to the trillion dollar electronics industry or priceless medical industry - ventures that can afford lofty goals with roadmaps. Of course one must beware of snake oil and asbestos, but is it really futile to imagine what new materials and markets might result from the current trends?

Tim: There are so many possibilities it's almost pointless to speculate on the killer applications, it's like medieval theologians debating how many angels can dance on the head of a pin—an interesting debate, but how do you verify your conclusions? The applications will surely come, but at the moment it's difficult to predict where from, simply because nanotechnology is a huge field and at a very early stage. We have identified several areas where we believe nanotechnology will have a significant, and early impact, but you have to keep an open mind, and stay in touch with the science, because things can change very rapidly, just ask any company who bought a 3G mobile telephony license!

Electricity, the Wheel and Nanotechnology

Steve: In which areas do you think nanotechnology will have a significant and early impact?

Tim: The recent Red Herring issue on Nanotechnology identified materials, medicine and optics. I would argue that the optical applications they mention are just exploiting properties of materials anyway, although I can understand their reasons for wanting to include optics, and in particular optical switching, in a magazine aimed at investors.

Nanotechnology is often called the new Internet, but its repercussions will be felt much more widely than that. I would liken it more to the invention of the telephone, or even the discovery of electricity. Now when we talk about future applications of nanotechnology, we should remember that electricity was initially used to replace gaslights & candles, the telephone was not designed to enable you to connect to the Internet. Put the technology in the hands of some bright people, and the results will be unimaginable, given our current perception of technology.

However, I would predict materials, electronics and pharmaceuticals are obvious areas in the short term, as a lot of the groundwork has already been done, and these industries already have the tools and experience of working at sub micron scales. While these may not be the most exciting applications of nanotechnology, they will improve existing materials & processes, and their commercialization will help fund a great deal of research into large scale manufacturing. Nanophase Technologies is a good example of how nanotechnology can have an impact in an unexpected way but this is just the tip of the iceberg. What's better, an army of nanobots mopping up skin cancer in twenty years time, or more efficient sun blocks now?

Steve: Perhaps progressive development and creative dreams are two complementary approaches to nanotechnology - like theory and experiment, top-down and bottom-up or even TNT and NTN. The wheel might be another technology worth remembering - seeing as for·so long we thought we had invented it, only to find that biomolecular nanotechnology has had wheels all along. In addition to faster computers, nanomedicine and mucho dinero, what are you looking forward to from nanotechnology?

Tim: Personally, I'm a big fan of any new technology, so of course I'm looking forward to all kinds of devices getting better, faster, lighter, and having less buttons. However, the area where nanotechnology has the greatest scope to change the world is by making things, whether they be mobile phones, computers, or drugs for the treatment of HIV or cancer, just work a whole lot better than they do at present.

REFERENCES

Albus, J.S.; Bainbridge, W.S.; Banfield, M.; Dastoor, C.A.; Murray, K.; Carley,M.; Hirshbein, T.; Masciangioli, T.; Miller, R.; Norwood, R.; Price, P.; Rubin, J.; Sargent,G.; Wallace, W.A.: 2003, 'Enhancing Group and Societal Outcomes: Theme D Summary', in M.C. Roco, W.S. Bainbridge (eds.), *Converging Technologies for Improving Human Performance*, Kluwer, Dordrecht, pp. 275-277.

Barber, Benjamin. 1996. *Jihad vs. McWorld: How Globalism and Tribalism are Reshaping the World* (New York: Ballantine Books).

Bijker, W. E. *Of Bicycles, Bakelites, and Bulbs: Toward a Theory of Sociotechnical Change*. Cambridge, MA: MIT Press, 1995.

Blumenthal, D.M., M. Gluck, K.S. Louis, M. Stoto, and D. Wise. 1986. "University-Industry Research Relationships in Biotechnology: Implications for the University," *Science* 232: pp. 1361-1366 (June 13).

Brave, Ralph. 2000. "Genome Hell is Revealed in Many Forms." *Baltimore Sun*, July 2, p. 1C.

Brown, J. S. and Duguid, P. Don't Count Society Out: A Response to Bill Joy. Paper prepared for the Societal Implications of Nanotechnology Conference, 2000.

Bruchez, M.; Moronne, M.; Gin, P.; Weiss, S.; Alivisatos, A.P. 1998. "Semiconductor nanocrystals as fluorescent biological labels", *Science*, 281, 2013-2016.

Casey, Jr., H.C., and M.B. Panish. 1978. *Heterostructure Lasers, Part A: Fundamental Principles*. (Academic Press. New York). pp. 1-9.

Cerf, Christopher, and Victor Navasky. 1984. *The Experts Speak*. New York: Pantheon Books.

Christensen, Clayton M. *The Innovator's Dilemma: When New Technologies Cause Great Firms to Fail*. New York: Harperbusiness, 2000.

Cipra, Barry A. 2000. The best of the 20th century: editors name top 10 algorithms. *SIAM News*. vol. 33(4).

Cook, T. K. and Campbell, D. T. *Quasi-Experimentation*. Rand-McNally, 1979 Fischer, C. S. *America Calling: A Social History of the Telephone to 1940*. Berkeley: U. of California, 1992.

Csicsery-Ronay, I. Jr.: 1991, 'The SF of Theory: Baudrillard and Haraway', *Science Fiction Studies*, 18, 387-404.

David, Paul. 1991. "Computer and Dynamo: The Modern Productivity Paradox in a Not-too-Distant Mirror," *Technology and Productivity: The Challenge for Economic Policy* (Paris: OECD).

DeSimone, J.M., E.T. Samulski, et al. 2000. Critical micelle density for the self-assembly of block copolymer surfactants in supercritical carbon dioxide. *Langmuir*. Vol.16. 416-421.

Di Paolo, E. A., *Artificial Life Bibliography of On-line Publications*, June 28, 2000.

Doering, Robert and Yoshio Nishi. 2000. "Limits of Integrated-Circuit Manufacturing." *Proceedings of the IEEE*. September.

Doi, M. and S.F. Edwards. 1986. *The Theory of Polymer Dynamics*. Oxford University Press (Clarendon), London-New York.

Donald, A.M. and A.H. Windle. 1992. Liquid crystalline polymers. *Cambridge Solid State Science Series*. Cambridge University Press.

Drell, S. D., A. D. Sofaaer, and G. D. Wilson, Hoover Digest 2000, No. 1, *The Present Threat*.

Dym, H. and H.P. McKean,. 1972. *Fourier Series and Integrals*. Academic Press.

Elkins, C.: 1979, 'Science Fiction versus Futurology: Dramatic Versus Rational Models', *Science Fiction Studies*, 6, 20-31.

Ellul, J. 1964. *The Technological Society*. New York: Alfred Knopf. (First published 1954).

Empedocles, S.; Bawendi, M. 1999. "Spectroscopy of single CdSe nanocrystallites," *Accounts of Chemical Research*, 32, 389-396.

Etzkowitz, H., M. Gulbrandsen and J. Levitt. 2000. *Public Venture Capital: Government Funding Sources for Technology Entrepreneurs*. New York: Harcourt.

Evans, J.H.: 2002, *Playing God? Human Genetic Engineering and the Rationalization of Public Bioethical Debate*, Chicago University Press, Chicago.

Fantel, Hans. 1987. "Portable CD Players Advance." *New York Times*, May 17, Section 2, p. 33, col. 1.

Fermi, E., J. Pasta, and S. Ulam. 1974. Studies of nonlinear problems, I. *Los Alamos Report*. LA1940. 1955.

Feynman, R.P. 1961. "There is plenty of room at the bottom." In *Miniaturization*. New York: Reinhold.

Fisher, D. E. and Fisher, M. J. *Tube: The Invention of Television*. Harcourt Brace, 1997.

FitzGerald, Frances. 2000. *Way Out There in the Blue: Reagan, Star Wars, and the End of the Cold War*. New York: Simon & Schuster.

Fogelberg, H.: 2003, '"Science" and "Technology" as Knowledge Cultures', in: H. Fogelberg, H. Glimell, *Bringing Visibility to the Invisible: Towards a Social Understanding of Nanotechnology*, Göteborg University, Göteborg, pp. 77-85.

Fogelberg, H.; Glimell, H.: 2003, *Bringing Visibility to the Invisible: Towards a Social Understanding of Nanotechnology*, Göteborg University, Göteborg.

Forest, G. 2000. Comment at Conference on Societal Implications of Nanoscience and Nanotechnology, September 29, 2000, NSF, Arlington, VA.

Forest, M.G., Q. Wang, and H. Zhou. 2000b. Homogeneous pattern selection and director instabilities of nematic liquid crystal polymers induced by elongational flows. *Physics of Fluids*. Vol. 12(3). 490-498.

Forum Proceedings of Sigma Xi, the Scientific Research Society, *Ethics, Values, and the Promise of Science*, Feb. 25-26, 1993.

Foucault, M. 1977. *Power/Knowledge: Selected Interviews and Other Writings 1972-1977*. New York, Pantheon Books.

Foucault, M.: 1992, *The Archaeology of Knowledge*, Routledge, London.

Fourier, J. 1878. *The Analytical Theory of Heat*. translated by A. Freeman. Cambridge University Press. London and New York.

Franklin, N.R.; Dai, H.J. 2000. "An enhanced CVD approach to extensive nanotube networks with directionality", *Advanced Materials*, 12, 890-894.

Frenken, J.W.M. 1998. "Scanning Tunneling Microscopy", section 5.2, A. ten Wolde (ed.), *Nanotechnology: Towards a molecular construction kit*. STT Report #60, The Hague, 289-299.

Gates, B.; Park, S.H.; Xia, Y.N. 2000. "Tuning the photonic bandgap properties of crystalline arrays of polystyrene beads by annealing at elevated temperatures", *Advanced Materials*, 12, 653-656.

Geneva Convention, *Protocol for the Prohibition of the Use in War of Asphyxiating, Poisonous or Other Gases, and of Bacteriological Methods of Warfare*, signed at Geneva on June 17,1925.

Gergen, K. J. Social psychology as history. *Journal of Personality and Social Psychology*, 1973, 26, 309- 20.

Gerlach, N.; Hamilton, S.: 2000, 'Telling the Future, Managing the Present: Business Restructuring Literature as SF', *Science Fiction Studies*, 27, 461-477.

Gieryn, T.: 1999, *Cultural Boundaries of Science*, Univ. of Chicago Press, Chicago.

Goldbeck-Wood, G. and A.H. Windle. 1999. Lattice modeling of nematodynamics. *Rheol. Acta*. Vol. 38. 548-561.

Graham, P.: 2002, 'Hypercapitalism: Language, New Media and Social Perceptions of Value', *Discourse and Society*, 13(2), 227-249.

Grosberg, A.Y. and A.R. Khokhlov. 1994. *Statistical Physics of Macromolecules*, AIP Series in Polymers and Complex Materials, AIP Press, New York.

Gubrud, M. A., *Nanotechnology and International Security*, Fifth Foresight Conference on Molecular Nanotechnology, November 5-8, 1997. Ref: http://www.foresight.org/Conferences/MNT05/Papers/Gubrud/.

Gurley, Bill. 2000. Digital music: the real law is Moore's law. *Fortune*. 2 October. page 268. Joy, Bill. 2000. Why the future doesn't need us. *Wired*. April. pp. 238 - 262.

Hayes, B., "The Square Root of NOT," *American Scientist*, July-August 1995, pp. 304-308.

Hayward, R.C.; Saville, D. A.; Aksay, I.A. 2000. "Electrophoretic assembly of colloidal crystals with optically tunable micropatterns", *Nature*, 404, 56-59.

Hendrickson, C.T., A. Horvath, S. Joshi, and L.B. Lave. 1998. Economic input-output models for environmental life-cycle assessment. *Environmental Science & Technology*. p. 32.

Hughes, P. M., *Global Threats and Challenges: The Decades Ahead*, Lt. Gen. Patrick M. Hughes, Director of the Defense Intelligence Agency, in Statement for the Senate Committee on Intelligence, January 28, 1998. Ref:http://www.fas.org/irp/congress/1998_hr/s980128h.htm.

James, E.: 1994, *Science Fiction in the 20th Century*, Oxford University Press, Oxford.

Jameson, F.: 1982, 'Progress Versus Utopia; or, Can We Imagine the Future', *Science Fiction Studies*, 9, 147-158.

Janet Raloff, "Nano Hazards: Exposure to minute particles harms lungs, circulatory system," Science News Online, Week of March 19, 2005; Vol. 167, No. 12.

Johnson, D.G. and H. Nissenbaum (eds). 1995. *Computers, Ethics and Social Values*. Prentice Hall:Englewood Cliffs, NJ. pp. 262-393.

Johnson, P.C.: 2003, 'Implications of the Continuum of Bioinformatics', in: M.C. Roco; W.S. Bainbridge (eds.), *Converging Technologies for Improving Human Performance*, Kluwer, Dordrecht, pp. 207-212.

Kalman B., F.D. Lublin. 1999. "The Genetics of Multiple Sclerosis. A Review." *Biomedicine and Pharmacotherapy*, Sep; 53 (8): pp. 358-370.

Kay, L.: 2000, *Who Wrote the Book of Life? A History of the Genetic Code*, Stanford University Press, Stanford.

Kruskal, M., and N. Zabusky. 1965. Interaction of solitons in a collisionless plasma and the recurrence of initial states, *Phys. Rev. Lett.* Vol. 15. 240-243.

Kupferman, R., M. Kawaguchi, and M.M. Denn. 2000. Emergence of structure in models of liquid crystalline polymers with elastic coupling. *J. Non-Newtonian Fluid Mech.* Vol. 91. 255-271.

Kuusi, O. 1999b. *Expertise in the future use of generic technologies*. VATT Research Reports #59, Helsinki.

Kuusi, O. and M. Meyer. 2001. "Technology Generalizations and Leitbilds – The Anticipation of Technological Opportunities." Manuscript submitted to *Technological Forecasting and Social Change*.

Kuusi, O., 1999. "Epistemic value analysis – a tool to evaluate technological options." Personal communication.

Laibinis, P.E.; Hickman, J.J.; Wrighton, M.S.; Whitesides, G.M. 1989. "Orthogonal Self-Assembled Monolayers - Alkanethiols On Gold and Alkane Carboxylic-Acids On Alumina", *Science*, 245, 845- 847.

Lambright, W. Henry. 1995. *Powering Apollo: James E. Webb of NASA*. Baltimore: Johns Hopkins University Press.

Latour, B. & Woolgar, S.: 1986, *Laboratory Life: The Social Construction of Scientific Facts*, 2nd edn., Princeton University Press, Princeton.

Latour, B.: 1986, 'Visualisation and Cognition: Thinking With Eyes and Hands', *Knowledge and Society*, 6, 1-40.

Lave, L.B., E. Cobas-Flores, C.T. Hendrickson, and F.C. McMichael. 1995a. Using input-output analysis to estimate economy wide discharges. *Environmental Science & Technology*, 29(9).

Lax, Peter D. 1989. The flowering of applied mathematics in America. in *A Century of Mathematics in America-Part II*. edited by Peter Duren et al. American Mathematical Society. 455-466.

Leah Eisenberg, "Medicating Death Row Inmates So They Qualify for Execution," AMA Case in Health Law, Vol. 6, No. 9, September 2004.

Ledley, Fred D. 1995. "Nonviral Gene Therapy: The promise of Genes as Pharmaceutical Products," *Human Gene Therapy*, Vol. 6, Sep, pp 1129-1144.

Lipowsky, R. 1991. The conformation of membranes. *Nature*. Vol. 349. 475.

Lipson, H. and J. B. Pollack, Nature, 406, 974-978, (2000). *Automatic design and manufacture of robotic life forms.* See, also, several additional articles on this subject in contemporary issues of Nature.

Mahmood U., C.H. Tung, A. Bogdanov Jr., R. Weissleder. 1999. "Near-Infrared Optical Imaging of Protease Activity for Tumor Detection." *Radiology*, Dec; 213 (3): pp. 866-870.

Malinow, R., Z.F. Mainen, and Y. Hayashi. 2000. LTP mechanisms: from silence to four-lane traffic. *Curr. Opin. Neurobiol*. 10(3): p. 352-7.

Mark, H., Statement of The Honorable Hans Mark, Director, Defense Research and Engineering, before the House Armed Service Committee, October 20, 1999.

McConnell, S.A., ed. 1998. *Pharmaceuticals*. Dun and Bradstreet Industry Reference Handbooks, Gàle, Detroit, pp. 30-34.

McHale, B.: 1987, *Postmodern Fiction*, Methuen, New York & London.

Meadows, D.L., et al. 1972. *The Limits to Growth*, New York: University.

Merz, J.L. 1986. "The Optoelectronics Joint Research Laboratory: Light Shed on Cooperative Research in Japan." *Scientific Bulletin* (ONR Far East Office). 11 (4), 1-30 (Oct.-Dec.).

Meyer, M. and O. Persson. 1998. "Nanotechnology - interdisciplinarity, patterns of collaboration and differences in application," *Scientometrics*, XXXXII (2), 195-205.

Milburn, C.: 2002 'Nanotechnology in the Age of Posthuman Engineering: Science Fiction as Science', *Configurations*, 10, 261-295.

Mnyusiwalla, A.; Daar, A.S.; Singer, P.A.: 2003, "'Mind the Gap": Science and Ethics in Nanotechnology, *Nanotechnology*, 14, R9-R13.

Nardi, B., O'Day, V. 1999. *Information Ecologies: Using Technology with Heart*. MIT Press.

National Science and Technology Council (NSTC). 2000. *National Nanotechnology Initiative: Leading to the Next Industrial Revolution* (Supplement to the President's FY 2001 Budget) (Washington, DC: February; available at http://www.nano.gov).

National Science and Technology Council (NSTC). 2000. *National Nanotechnology Initiative: Leading to the Next Industrial Revolution*. February. (http://www.nano.gov/nni.pdf).

Parikh, A.N.; Schivley, M.A.; Koo, E.; Seshadri, K.; Aurentz, D.; Mueller, K.; Allara, D.L. 1997. "nalkylsiloxanes: From single monolayers to layered crystals. The formation of crystalline polymers from the hydrolysis of n- octadecyltrichlorosilane", *Journal of the American Chemical Society*, 119, 3135-3143.

Philip Ball, "Synthetic Biology: Starting from Scratch," Nature, 431, pp. 624-626, 7 October 2004.

Philipse, A.P. 1998. "Colloidal Dispersions," section 3.4, A. ten Wolde (ed.), *Nanotechnology: Towards a molecular construction kit*. STT Report #60, The Hague, 171-178.

Pierce, N.A. and M.B. Giles. 2000. Adjoint reconstruction of superconvergent functionals from PDE approximations. *SIAM Rev.* Vol. 42(2). 247-264.

Piveteau, Laurent-Dominique, Anthe S. Zandvliet and Robert Langer. 2001. "Nitroxide and Galactose Grafted Dendrimer: A Specifically Targeted Contrast Agent for MRI and EPR Imaging of the Liver," *in preparation.*

Polanyi, Karl. 1944. *The Great Transformation: The Political and Economic Origins of Our Time* (Boston: Beacon Press, 1957; orig. ed. 1944).

Pritchard, M.S. 1996. "Conflicts of Interest: Conceptual and Normative Issues," *Academic Medicine*, 71:12, December, pp. 1305-1313.

Putnam, D., Gentry, C.A., Pack, D.W., Langer, R. 2001. "Polymer-based Gene Delivery with Low Cytotoxicity by a Unique Balance of Side-chain Termini." *Proceedings of the National Academy of Sciences*, U.S.A., Volume 98, Number 3, pages 1200-1205.

Ramos, L.; Lubensky, T.C.; Dan, N.; Nelson, P.; Weitz, D.A. 1999. "Surfactant-mediated two-dimensional crystallization of colloidal crystals", *Science*, 286, 2325-2328.

Reed, Mark A. and James M. Tour. 2000. "Computing with Molecules." *Scientific American*. June.. SIA (Semiconductor Industry Association). 1999. "1999 International Technology Roadmap for Semiconductors."

Rinzler, A.G.; Liu, J.; Dai, H.; Nikolaev, P.; Huffman, C.B.; Rodriguez-Macias, F.J.; Boul, P.J.; Lu, A.H.; Heymann, D.; Colbert, D.T.; Lee, R.S.; Fischer, J.E.; Rao, A.M.; Eklund, P.C.; Smalley, R.E. 1998.

Roco, M. C., R. S. Williams, and P. Alivisatos, *Nanotechnology Research Directions: Vision for Nanotechnology in the Next Decade*, Kluwer Academic Publishers (2000).

Roco, M.C. & Bainbridge, W.S., (eds.): 2001, *Societal Implications of Nanoscience and Nanotechnology*, National Science Foundation.

Roco, M.C.: 2003b, 'Coherence and Divergence in Megatrends in Science and Engineering', in: M.C. Roco, W.S. Bainbridge (eds.), *Converging Technologies for Improving Human Performance*, Kluwer, Dordrecht, pp. 79-96.

Roco, M.C.: 2003c, 'Fundamentally New Manufacturing Processes and Products', in: Roco, M.C. & Bainbridge, W.S., (eds.), *Converging Technologies for Improving Human Performance*, Kluwer, Dordrecht, pp. 300-302.

Roco, M.C.: 2004, 'The Emergence and Policy Implications of Converging Technologies' [www.nsf.gov/home/crssprgm/nano/nbic_roco_04_0422_@aaas_57sl.pdf, accessed May 2004].

Rodney Brooks, "The Merger of Flesh and Machines," The Next Fifty Years: Science in the First Half of the Twenty-First Century, ed. by John Brockman, 2002, p. 191.

Roland Pease, "'Living' robots powered by muscle," BBC News, January 17, 2005.

Schummer, J.: 2004, 'Interdisciplinary Issues in Nanoscale Research', in: D. Baird, A. Nordmann, J. Schummer (eds.), *Discovering the Nanoscale*, IOS Press, Amsterdam, pp. 9-20.

Schumpeter, Joseph Alois. 1942. *Capitalism, Socialism, and Democracy*. New York: Harper.

Senturia, S., N. Aluru, and J. White. 1997. Simulating the behavior of MEMS devices: computational methods and needs. *IEEE Computational Science and Engineering*. Vol. 4(1). 30-43.

Shor, P., *Proc. 35th Ann. Symp. Foundations of Computer Science*, 1994.

Shultz, W.: 2000, 'Nanotechnology: The Next Big Thing', *Chemical and Engineering News*, 78(18), 41-47.

Snider, G.L.; Orlov, A.O.; Amlani, I.; Zuo, X.; Bernstein, G.H.; Lent, C.S.; Merz, J.L.; Porod, W. 1999. "Quantum-dot cellular automata: Review and recent experiments (invited)", *Journal of Applied Physics*, 85, 4283-4285.

Soong, R.K.; Bachand, G.D.; Neves, H.P.; Olkhovets, A.G.; Craighead, H.G.; Montemagno, C.D. 2000. "Powering an inorganic nanodevice with a biomolecular motor", *Science*, 290, 1555-1558.

Steelman, John R. 1947. *Science and Public Policy* (Washington, DC: US Government Printing Office). Stephenson, Neal. 1995. *The Diamond Age, or, Young Lady's Illustrated Primer* (New York: Bantam Books).

Stix, G. 1996. Trends in nanotechnology: waiting for breakthroughs. *Sci. Am.* April.

Sun, S.H.; Murray, C.B.; Weller, D.; Folks, L.; Moser, A. 2000. "Monodisperse FePt nanoparticles and ferromagnetic FePt nanocrystal superlattices", *Science*, 287, 1989-1992.

Suvin, D.: 1979, *Metamorphoses of Science Fiction*, Yale University Press, New Haven.

Tilton, John E. (ed). 1991. *World Metal Demand*, Washington, DC: Resources for the Future.

Tobio, M., R. Gref, A. Sanchez, R. Langer and M.J. Alonso. 1998. "Stealth PLA-PEG Nanoparticles as Protein Carriers for Nasal Administration", *Pharmaceutical Research*, Vol. 15, No 2, pp 270-275.

Tramontin, A.D. and E.A. Brenowitz. 2000. Seasonal plasticity in the adult brain. *Trends Neurosci.* 23(6): p. 251-8.

Vaia, R., D. Tomlin, M. Schulte, and T. Bunning. 2000. Two-phase nanoscale morphology of polymer/LC composites. *Polymer*. In press.

Vale, R.D.; Milligan, R.A. 2000. "The way things move: Looking under the hood of molecular motor proteins", *Science*, 288, 88-95.

Valenti. Plaza y Valdes, S.A. de C.V. Instituto de Investigaciones Sociales. Universidad Nacional Autonoma de Mexico. Universidad Autonoma Metropolitana, pp.81-93.

Van Maanen, J. *Tales of the Field*. Chicago: U. Chicago, 1988.

Vettiger, P.; Brugger, J.; Despont, M.; Drechsler, U.; Durig, U.; Haberle, W.; Lutwyche, M.; Rothuizen, H.; Stutz, R.; Widmer, R.; Binnig, G. 1999. "Ultrahigh density, high-data-rate NEMS-based AFM data storage system", *Microelectronic Engineering*, 46, 11-17.

Voter, A.F. 1997. Hyperdynamics: accelerated molecular dynamics of infrequent events. *Phys. Rev. Lett.* Vol. 78. 3908.

Warner, M. and E. M. Terentjev. 1996. Nematic elastomers: a new state of matter? *Progress in Polymer Science*. Vol. 21(5). 853.

Warner, M., E. M. Terentjev, R. B. Meyer, and Y. Mao. 2000. Untwisting of a cholesteric elastomer by a mechanical field. *Phys. Rev. Lett.* Vol. 85(11). 2320.

Webb, E. J., Campbell, D. T., Schwartz, R. D., Sechrest, L., and Grove, J. B. *Unobtrusive Measures: Nonreactive Research in the social Sciences* (2nd ed.). Boston: Houghton Mifflin, 1981.

Webster, A. and H. Etzkowitz. 1991. "Academic-industry Relations: The Second Academic Revolution?" Science Policy Support Group, London, 1991.

Weil, V. 1988. "Afterword," *Biotechnology: Professional Issues and Social Concerns*. (AAAS Publication) co-edited with Paul DeForest, Mark Frankel, and Jeanne Poindexter.

Weiss, S. 1999. "Fluorescence spectroscopy of single biomolecules", *Science*, 283, 1676-1683.

White, J.G. 1986. *Philos. Trans. R. Soc. Lond. B Biol. Sci.* 314: pp. 1-340.

Whitesides, G.M.; Mathias, J.P.; Seto, C.T. 1991. "Molecular Self-Assembly and Nanochemistry - a Chemical Strategy For the Synthesis of Nanostructures", *Science*, 254, 1312-1319.

Wilbur, J.L.; Whitesides, G.M. 1999. *Self-Assembly and Self-Assembled Monolayers in Micro- and Nanofabrication*; Timp, G., Ed.; Springer-Verlag New York, Inc.: New York, pp 331-370.

William-Jones, B.; Corrigan, O.: 2003, 'Rhetoric and Hype: Where's the Ethics in Pharmacogenomics?', *American Journal of Pharmacogenomics*, 3(6), 375-383.

3

Recent Developments and News Shaping Future Direction of Nano Labs

X-rays and Neutrons: Essential Tools for Nanoscience Research

Virtually all of the grand challenges in nanotechnology have as overriding themes the need to determine the structure over a wide range of length scales and to understand how the combination of the structure and dynamics leads to their unique properties. Both x-rays and neutrons cover all the length scales of interest in nanoscience, from the atomic structure of individual building blocks to the configuration of assembled, functional structures, making them essential tools for the elucidation of these challenges.

The workshop is one of a series of workshops in support of the National Nanotechnology Initiative (NNI) Strategic Plan prepared by the Nanoscale Science and Engineering and Technology Subcommittee (NSTC), recently updated in December 2004 as part of the 21st Century Nanotechnology Research and Development Act (Public Law 108-153).

As called for in the Strategic Plan, the workshop will provide a guiding strategy to policy makers for the development of essential scattering tools for R&D in nanotechnology. The workshop report will be developed by internationally renowned participants who will identify frontier problems in their fields from experimental and theoretical perspectives by addressing the essential questions:

- Which of the outstanding problems in nanoscale synthesis, structure, dynamics and properties can be addressed using X-ray and neutron techniques such as scattering, imaging and spectroscopy, and how can these techniques help illuminate the important and urgent issues at the nanoscale?

- How might the current resource base in X-ray and neutron scattering techniques be augmented and used in solving outstanding problems in nanoscale science?

A scholarship program will also provide the opportunity for selected early career researchers to participate fully in the workshop and the development of the ensuing report.

Projected Outcomes

The workshop will bring together established experts and young scientists from the x-ray, neutron, and nanoscience/nanotechnology communities to discuss future research directions and opportunities. The goal of the workshop is to identify and promote the innovations in scattering techniques and related instrumentation required to tackle the exciting challenges of nanotechnology research now and for the future. The focus of this workshop will be to identify frontier problems in nanoscience such as understanding interfacial structures, nano-systems, confinement, and self-assembly of hard materials, soft materials, and biomaterials which can be elucidated through the use of x-ray and neutron scattering techniques.

After these research opportunities are identified, the workshop report will outline a roadmap for the prioritized development of these critical resources. Specifically, the workshop report will address the following goals:

- Identify the challenges in characterization at the nanoscale for the next 5-10 years that may be addressed using x-ray or neutron scattering techniques.
- Identify the instrumentation and techniques that must be developed to meet these challenges and allow research in nanoscience to advance.
- Identify both short and long term R&D in areas such as beam optics, detectors and in-situ characterization that will be required to support this vision.

The workshop report will be an effective tool for funding agencies, principal investigators, academia and industry to coordinate their long term investment strategies.

Supported by

- U.S. Department of Energy, Office of Basic Energy Sciences
- National Institute of Standards and Technology, NIST Center for Neutron Research
- National Science Foundation, Division of Materials Sciences

Workshop Chairs

- Ian Anderson (Oak Ridge National Laboratory, Spallation Neutron Source)

- Linda Horton (Oak Ridge National Laboratory, Center for Nanophase Materials Sciences)
- Eric Isaacs (Argonne National Laboratory, Center for Nanoscale Materials, and University of Chicago)
- Mark Ratner (Northwestern University)

Workshop Organizers:

- Ian Anderson (Oak Ridge National Laboratory, Spallation Neutron Source)
- Kristin Bennett (U.S. Department of Energy, Office of Basic Energy Sciences)
- Al Ekkebus (Oak Ridge National Laboratory, Spallation Neutron Source)
- Pat Gallagher (National Institute of Standards and Technology)
- Phillip Lippel (National Nanotechnology Coordination Office)
- Linda Horton (Oak Ridge National Laboratory, Center for Nanophase Materials Sciences)
- Eric Isaacs (Argonne National Laboratory, Center for Nanoscale Materials)
- Helen Kerch (U.S. Department of Energy, Office of Basic Energy Sciences)
- Celia Merzbacher (Office of Science and Technology Policy)
- Mark Ratner (Northwestern University)
- Guebre Tessema (National Science Foundation, Division of Materials Research)

The Nanosurf® Mobile S

The Mobile S atomic force microscope is a whole new way to look at atomic force microscopy. Mobility. Ease of use. Sleek design. Modern and well-integrated software. Unsurpassed support and reliability. And a price much lower than you'd pay for a comparable AFM system.

These are features that cannot be matched by any other AFM. This revolution in design increases your productivity, decreases your downtime, and gives you more time to focus on your real application issues. Being an expert in AFM is no longer required to get results. The mobile S features the most commonly used modes of AFM operation. Dynamic Force or "tapping mode", contact mode, force modulation, phase contrast imaging, surface spreading resistance, and force spectroscopy are among the supported modes. Additionally, user inputs and customized scripting is supported with all modes. Integrated Dual-View cameras, elimination of laser and detector alignment, and intuitive software sets the Mobile S apart from all other atomic force microscope systems.

Nanosceince Instruments

Nanoscience Instruments is your one source for a wide variety of AFM systems, probes, accessories, and related nanoscience tools. Our products include the easy-to-use Nanosurf® easyScan SPMs, to the advanced WITec AlphaSNOM systems, as well as a wide variety of AFM probes, AFM add-on accessories, and post processing software for virtually every SPM system.

Tools for Nano-technology

Electron Microscopy

In 1924 L. de BROGLIE discovered the wave-character of electron rays thus giving the prerequisite for the construction of the electron microscope. The prototype was built by M. KNOLL and E. RUSKA (Technische Universität Berlin, 1932). One of the first biological objects depicted was the tobacco mosaic virus (TMV). The first picture of a cell was published in 1945 by K. R. PORTER, A. CLAUDE and E. F. FULLAM (Rockefeller Institute, New York).

The Transmission Electron Microscope (TEM)

The conventional electron microscopy is nowadays called TEM (transmission electron microscopy). We will therefore start with its construction. The ray of electrons is produced by a pin-shaped cathode heated up by current. The electrons are vacuumed up by a high voltage at the anode. The acceleration voltage is between 50 and 150 kV. The higher it is, the shorter are the electron waves and the higher is the power of resolution. But this factor is hardly ever limiting. The power of resolution of electron microscopy is usually restrained by the quality of the lens-systems and especially by the technique with which the preparation has been achieved. Modern gadgets have powers of resolution that range from 0,2 - 0,3 nm. The useful resolution is therefore around 300,000 x.

The accelerated ray of electrons passes a drill-hole at the bottom of the anode. Its following way is analogous to that of a ray of light in a light microscope. The lens-systems consist of electronic coils generating an electromagnetic field. The ray is first focused by a condenser. It then passes through the object, where it is partially deflected. The degree of deflection depends on the electron density of the object. The greater the mass of the atoms, the greater is the degree of deflection. Biological objects have only weak contrasts since they consist mainly of atoms with low atomic numbers (C, H, N, O). Consequently it is necessary to treat the preparations with special contrast enhancing chemicals (heavy metals) to get at least some contrast. Additionally they are not to be thicker than 100 nm, because the temperature is raising due to electron absorption. This again can lead to destruction of the preparation. It is generally impossible to examine living objects.

After passing the object the scattered electrons are collected by an objective. Thereby an image is formed, that is subsequently enlarged by an additional lens-system (called projective with electron microscopes). The thus formed image is made visible on a fluorescent screen or it is documented on photographic material. Photos taken with electron microscopes are always black and white. The degree of darkness corresponds to the electron density (= differences in atom masses) of the candled preparation.

The Scanning electron microscope (SEM)

The path of the electron beam within the scanning electron microscope differs from that of the TEM. The technology used is based on television techniques. The method is suitable for the depiction of preparations with conductive surfaces. Biological objects have thus to be made conductive by coating with a thin layer of heavy metal (usually gold is taken). The power of resolution is normally smaller than in transmission electron microscopes, but the depth of focus is several orders of magnitude greater. Scanning electron microscopy is therefore also well-suited for very low magnifications. Numerous examples will be given in the following.

The surface of the object is scanned with the electron beam point by point whereby secondary electrons are set free. The intensity of this secondary radiation is dependent on the angle of inclination of the object's surface. The secondary electrons are collected by a detector that sits at an angle at the side above the object. The signal is then enhanced electronically. The magnification can be chosen smoothly (depending on the model) and the image appears a little later on a viewing screen.

The properties of the light microscope as opposed to that of transmission and scanning electron microscopes are collected in a table.

Finally some outlines of new and further developments are given. The high voltage electron microscope: it operates with an accelerating voltage of 700 - 3000 kV. Its power of resolution is greater, the preparation can be thicker, the strain on the preparation is smaller. But the enormous technical expenditure is disadvantageous. Only few gadgets exist. New results concerning botany have not been gained. The scanning transmission electron microscope (STEM): In this development of the SEM do the electrons pass through the preparation and the secondary radiation thus generated is used for image formation. Here, too, the expenditure is large, but it is still worthwhile, since large molecules like nucleic acids or proteins or molecular complexes like viruses can be depicted much better and gentler than with the TEM. No news for botany, though. The interpretation of images gained with electron microscopy is increasingly done with computerized interpretation programs. But they are usually only suitable for the reconstruction of regularly recurring patterns and these, again, are found more often on a molecular level than on a cellular one.

Atomic Force Microscopy—AFM tools

The Company: Since the late 1990's Novascan has been a nanotechnology company that specializes in atomic force microscope products (AFM) and AFM / SPM services. Our tools are used worldwide to visualize, characterize and manipulate microscopic and nanoscopic environments.

Our Mission: We are committed to developing and producing innnovative and state of the art devices that facilitate our clients projects in nanotechnology. These improvements range from specialized AFM instruments to the probes and sensors used to characterize samples.

Technology: At Novascan we strive to develop forward looking devices with the hope that we can help you make new scientific discoveries in your work and projects. Want to learn more?

Scanning Tunneling Microscopy

Quantum Corrals

Scientists discovered a new method for confining electrons to artificial structures at the nanometer lengthscale. Surface state electrons on Cu(111) were confined to closed structures (corrals) defined by barriers built from Fe adatoms. The barriers were assembled by individually positioning Fe adatoms using the tip of a low temperature scanning tunneling microscope (STM). A circular corral of radius 71.3 Angstrom was constructed in this way out of 48 Fe adatoms. STM image shows the direct observation of standing-wave patterns in the local density of states of the Cu(111) surface. These spatial oscillations are quantum-mechanical interference patterns caused by scattering of the two-dimensional electron gas off the Fe adatoms and point defects.

History of the Microscope

Light Microscopes, Electron Microscopes, Scanning Electron Microscope

. During that historic period known as the Renaissance, after the "dark" Middle Ages, there occurred the inventions of printing, gunpowder and the mariner's compass, followed by the discovery of America. Equally remarkable was the invention of the microscope: an instrument that enables the human eye, by means of a lens or combinations of lenses, to observe enlarged images of tiny objects. It made visible the fascinating details of worlds within worlds.

Long before, in the hazy unrecorded past, someone picked up a piece of transparent crystal thicker in the middle than at the edges, looked through it, and discovered that it made things look larger. Someone also found that such a crystal would focus the sun's rays and set fire to a piece of parchment or cloth. Magnifiers and "burning glasses" or "magnifying glasses" are mentioned in the writings of Seneca and Pliny the

Elder, Roman philosophers during the first century A. D., but apparently they were not used much until the invention of spectacles, toward the end of the 13th century. They were named lenses because they are shaped like the seeds of a lentil.

The earliest simple microscope was merely a tube with a plate for the object at one end and, at the other, a lens which gave a magnification less than ten diameters — ten times the actual size. These excited general wonder when used to view fleas or tiny creeping things and so were dubbed "flea glasses. "

About 1590, two Dutch spectacle makers, Zaccharias Janssen and his son Hans, while experimenting with several lenses in a tube, discovered that nearby objects appeared greatly enlarged. That was the forerunner of the compound microscope and of the telescope. In 1609, Galileo, father of modern physics and astronomy, heard of these early experiments, worked out the principles of lenses, and made a much better instrument with a focusing device.

The father of microscopy, Anton van Leeuwenhoek of Holland (1632-1723), started as an apprentice in a dry goods store where magnifying glasses were used to count the threads in cloth. He taught himself new methods for grinding and polishing tiny lenses of great curvature which gave magnifications up to 270 diameters, the finest known at that time. These led to the building of his microscopes and the biological discoveries for which he is famous. He was the first to see and describe bacteria, yeast plants, the teeming life in a drop of water, and the circulation of blood corpuscles in capillaries. During a long life he used his lenses to make pioneer studies on an extraordinary variety of things, both living and non living, and reported his findings in over a hundred letters to the Royal Society of England and the French Academy.

Robert Hooke, the English father of microscopy, re-confirmed Anton van Leeuwenhoek's discoveries of the existence of tiny living organisms in a drop of water. Hooke made a copy of Leeuwenhoek's microscope and then improved upon his design.

Later, few major improvements were made until the middle of the 19th century. Then several European countries began to manufacture fine optical equipment but none finer than the marvelous instruments built by the American, Charles A. Spencer, and the industry he founded. Present day instruments, changed but little, give magnifications up to 1250 diameters with ordinary light and up to 5000 with blue light.

A light microscope, even one with perfect lenses and perfect illumination, simply cannot be used to distinguish objects that are smaller than half the wavelength of light. White light has an average wavelength of 0.55 micrometers, half of which is 0.275 micrometers. (One micrometer is a thousandth of a millimeter, and there are about 25,000 micrometers to an inch. Micrometers are also called microns.) Any two lines that are closer together than 0.275 micrometers will be seen as a single line, and any object with a diameter smaller than 0.275 micrometers will be invisible or, at best, show up as a blur. To see tiny particles under a microscope, scientists must bypass light altogether and use a different sort of "illumination," one with a shorter wavelength.

The introduction of the electron microscope in the 1930's filled the bill. Co-invented by Germans, Max Knott and Ernst Ruska in 1931, Ernst Ruska was awarded half of the Nobel Prize for Physics in 1986 for his invention. (The other half of the Nobel Prize was divided between Heinrich Rohrer and Gerd Binnig for the STM.) In this kind of microscope, electrons are speeded up in a vacuum until their wavelength is extremely short, only one hundred-thousandth that of white light. Beams of these fast-moving electrons are focused on a cell sample and are absorbed or scattered by the cell's parts so as to form an image on an electron-sensitive photographic plate.

If pushed to the limit, electron microscopes can make it possible to view objects as small as the diameter of an atom. Most electron microscopes used to study biological material can "see" down to about 10 angstroms—an incredible feat, for although this does not make atoms visible, it does allow researchers to distinguish individual molecules of biological importance. In effect, it can magnify objects up to 1 million times. Nevertheless, all electron microscopes suffer from a serious drawback. Since no living specimen can survive under their high vacuum, they cannot show the ever-changing movements that characterize a living cell.

Using an instrument the size of his palm, Anton van Leeuwenhoek was able to study the movements of one-celled organisms. Modern descendants of van Leeuwenhoek's light microscope can be over 6 feet tall, but they continue to be indispensable to cell biologists because, unlike electron microscopes, light microscopes enable the user to see living cells in action. The primary challenge for light microscopists since van Leeuwenhoek's time has been to enhance the contrast between pale cells and their paler surroundings so that cell structures and movement can be seen more easily. To do this they have devised ingenious strategies involving video cameras, polarized light, digitizing computers, and other techniques that are yielding vast improvements in contrast, fueling a renaissance in light microscopy.

International Federation of Societies for Microscopy

Membership List

As at January 2003, organised alphabetically. Click on the name of a country listed below to view their IFSM membership contact details. If the country you are searching for is not listed here, please look under either Associate Members or Regional Members.

- America (USA)
- Armenia
- Austria
- Australia

- Belgium
- Brazil
- Britain
- Canada
- China
- Columbia
- Croatia
- Czech and Slovak Republics
- France
- Germany
- Hungary
- India
- Ireland
- Israel
- Italy
- Japan
- Korea
- Mexico
- Netherlands
- New Zealand
- Portugal
- Poland
- Russia
- Scandinavia
- Singapore
- Slovenia
- Southern Africa
- Spain
- Switzerland
- Taipei China
- Thailand
- Turkey

- UK
- USA
- Venezuela

Microscopy Society of America (USA)

- **Dr. M. G. (Grace) Burke (President)**
  Dr. W. G. (Jay) Jerome **(President-Elect)**
  Business Office
  Judy Janes, Administrator
  Philip Lesser, Managing Director
  Bostrom Corporation
  230 E. Ohio Street, Suite 400
  Chicago, IL 60611-3265
  Tel: (800) 538-3672
  Fax: (312) 644-8557
  E-mail: businessoffice@microscopy.org
  http: //www.msa.microscopy.com/
- **Armenian Electron Microscopy Society**
  Att: Dr Karlen Hovnanyan
  Institute of Molecular Biology NAS Republic Armenia
  7 Hasratian St.
  Yerevan 375014
  Armenia
  Phone: +374 1 281 626,+ 374 1 550 209 ap.
  Fax: +374 1 544 015
  E.mail: hovkarl@freenet.am or hovaems@ngoc.am
- **Austrian Society for Electron Microscopy**
  Univ.Prof. Dr. Margit PAVELKA
  Medizinische Universität Wien
  Zentrum für Anatomie und Zellbiologie
  Institut für Histologie und Embryologie
  Abteilung für Zellbiologie und Ultrastrukturforschung
  Schwarzspanierstraße 17, A-1090 Wien

Tel: (+43-1) 4277-61361

Fax: (+43-1) 4277-61309

E-Mail: Margit.Pavelka@meduniwien.ac.at

http: //www.univie.ac.at/asem/

- **Australian Microscopy and Microanalysis Society Inc.**

  Dr. Kath Smith, President

  Dr Vicki Keast, Secretary

  Dr. Brendan Griffin, International Liaison Officer

  Australian Key Centre for Microscopy and Microanalysis

  University of Sydney, NSW 2006

  Australia

  Tel: +61 2 9351 4523

  Fax: +61 2 9351 7682

  Email (Secretary): mailto: amms.secretary@emu.usyd.edu.au

  Email (ILO): mailto: %20bjg@cmm.uwa.edu.au

  http: //www.microscopy.org.au/

- **Belgian Society for Microscopy**

  Prof Dr Dirk Adriaensen (Secretary)

  Department of Biological Sciences

  University of Antwerp

  Campus Middelheim

  Groenenborgerlaan 171

  B-2020 Antwerp

  Belgium

  Tel: +32-3-2653475

  Fax: +32-3-2653301

  Email: dirk.adriaensen@ua.ac.be

  http: //webhost.ua.ac.be/BVM_SBM/

- **Brazilian Society for Microscopy and Microanalysis**

  Prof Fernando Galembeck (Unicamp), President

  Prof Marcos André Vannier (Fiocruz/BA), Vice-President Biology

  Prof Carlos da Costa Viana (IME/RJ), Vice-President Materials

Prof Maria do Carmo Gonçalves (Unicamp), Secretary
Prof Carlos Alberto Paula Leite (Unicamp), Treasurer
Sociedade Brasileira de Microscopia e Microanálise
Instituto de Química—Unicamp – Campinas—SP
Caixa Postal: 6154
CEP: 13084-971
Brazil
Tel: +19 3788-3080
Fax: +19 3788-3080
Home Page: http: //www.sbmm.org.br/
Email: sbmm@iqm.unicamp.br

- **Royal Microscopical Society**
  Dr Chris Hawes, President
  37/38 St Clements
  Oxford
  OX4 1AJ
  England
  Tel: +44 01865 248768
  Fax: +44 01865 791237
  Email: info@rms.org.uk
  http: //www.ifsm.umn.edu/members.html#
- **Microscopical Society of Canada**
  Dr Pierre-M Charest
  Département de phytologie
  FSAA, Pavillon Paul-Comtois
  Universite Laval
  Québec
  CANADA G1K 7P4
  Tel: +1 418 656 7792
  Fax: +1 418 656 7856
  Email: Pierre-Mathieu.Charest@plg.ulaval.ca
  http: //msc.rsvs.ulaval.ca/

- **Chinese Electron Microscopy Society**
  Professor Heng-qiang Ye
  President
  Laboratory of Atomic Imaging of Solids
  Institute of Metal Research
  Chinese Academy of Sciences
  Shenyang 110016
  China
  Home Page: http: //www.cems.synl.ac.cn/Email: mailto: %20hgye@imr.ac.cn
- **Columbian Society of Electron Microscopy**
  Dr Maria Leonor Caldas
  President
  Microscopy and Imagen Analysis Group
  National Institute of Health
  Avenida Calle 26 No. 51-60
  Bogotá,
  Columbia
  Tel: +22 07700 ext. 453
  Fax: +22 21093
  Email: mcaldas@hemagogus.ins.gov.co
- **Croatian Society for Electron Microscopy**
  Prof Dr Djuro Vranesic
  Frankopanska 1/dvor
  HR-10000 Zagreb
  Croatia
  Email: mailto: djvranesic@vodatel.net
  http: //www.ifsm.umn.edu/members.html#
- **Czechoslovak Microscopy Society (CSMS)**
  Prof Pavel Hozak (President)
  c/o Institute of Experimental Medicine
  Videnska 1083
  142 20 Prague 4

Czech Republic
Tel/Fax: +420 241 062 219
Email: hozak@biomed.cas.cz
http: //www.microscopy.cz/

- **Société Francaise des Microscopies**
  Attn: Prof Genevieve Nihoul
  University of Paris VI
  Case 243 (Bat C, 6e étage)
  9 Quai Saint Bernard
  75252 Paris Cedex 05
  FRANCE
  Tel: +33 1 4427 2621
  Fax: +33 1 4427 2622
  Email: mailto: %20sfme@snv.jussieu.fr
  http: //sfmu.snv.jussieu.fr/
- **Deutsche Gesellschaft fuer Elektronenmikroskopie**
  Professor Dr Hannes Lichte
  Institut fuer Angewandte Physik
  TU Dresden, Triebenberglabor
  D-01062 Dresden
  Germany
  Tel: +49 351 21508910
  Fax: +49 351 21508920
  Email: hannes.lichte@triebenberg.de
  http: //www.dge-homepage.de/
- **Dutch Society for Microscopy**
  Prof. Dr. H.K. Koerten
  Molecular Cell Biology
  LUMC
  Postbus 9503
  2300 RA
  Leiden

The Netherlands
Fax: (+31) 71 527 64 40
Tel: (+31) 71 527 64 58
Email: h.k.koerten@lumc.nl

- **Dr JJL van der Want (Treasurer)**
  Department of Cell Biology Faculty of Medicine
  University of Groningen A Deusinglaan 1 (Bldg 3215, 7th & 11th Floor)
  9713 AV Groningen
  The Netherlands
  Tel: +31 50 3632522 / 31 50 3632501
  Fax: +31 50 3632512 / 31 50 3632515
  http: //www.microscopie.nl/
- **Hungarian Society for Microscopy**
  Dr Kristof Kovacs
  President, Hungarian Society for Microscopy
  University of Veszprem, P.O.Box 158, Veszprem
  H-8201 Hungary
  Tel: +36 88 421 684
  Fax: +36 88 328 643
  Email: mailto: %20kris@almos.vein.hu
  http: //picasso.elte.hu/microscopy
- **Electron Microscope Society of India**
  President:
  Prof G N Mathur
  Director of Defence Materials & Stores Research &
  Development Establishment
  PO: DMSRDE
  GT Road
  Kanpur 208013
  India
  Tel: 00-91-0512-450695/451759
  Res: 00-91-0512-359597

Fax: 00-91-0512-450404

mailto: dmsrde@sanchamet.in

Vice-Presidents:

Dr J Kumar (IIT, Kanpur

Mr ML Sharma (Punjab University, Chandigarh)

Hon. General Secretary:

Dr RB Srivastava—Joint Director

Joint Secretary:

Mr KN Pandey (DMSRDE, Kanpur) mailto: dmsrde@sanchamet.in

Treasurer:

Mr SB Yadaw (DMSRDE, Kanpur)

- **Microscopical Society of Ireland**

  Mr Alexander Black (President)

  Department of Anatomy

  National University of Ireland

  Galway

  Ireland

  Email: alexander.black@nuigalway.ie

  http: //www.ifsm.umn.edu/members.html#

- **Dr. Michael Ball (Honorary Secretary)**

  National Centre for Biomedical Engineering Science

  National University of Ireland,

  Galway,

  Ireland

  Tel: +353 91 524411 ext: 3780

  Email: michael.ball@nuigalway.ie

- **Dr Gerard Brennan (Treasurer)**

  School of Biology & Biochemistry The Queen's University of Belfast

  Belfast BT9 7BL

  Northern Ireland

- **Israel Society for Microscopy**

  Prof Ilan Hammel

Department of Pathology
Sackler Faculty of Medicine
Tel Aviv University
Ramat Aviv 69978
Israel
Phone: 972 3 640 8408(lab)/6159/9861
Fax: 972 3 640 9141
E-mail: ilanh@patholog.tau.ac.il
http: //www.ifsm.umn.edu/members.html#

- **Italian Society for Microscopical Sciences (SISM)**
Prof Daniela Quaglino (President)
Department of Biomedical Sciences—Genreal Pathology
Via Campi 287
41100 Modena
Italy
Tel: +0039 059 205 5442
Fax: +0039 059 205 5426
Email: quaglino.daniela@unimore.it
http: //www.sism.unile.it/

- **Japanese Society of Microscopy**
Akira Tonomura (President)
Ayano Satomi (General Secretary)
Ohtsuka 3cho-me Bldg.,3-11-6
Ohtsuka,Bunkyo-ku,
Tokyo 112-0012
Japan
Tel: +81 (0)3 5940 7640
Fax: +81 (0)3 5940 7980
email: kenbikyo@r-sipec.jp

- **Korean Society of Electron Microscopy**
Dong-Wha Kum
Division of Metals Room 0351

Korea Institute of Science and Technology (KIST)
Cheongryang P.O. Box 131
Seoul, 136-791
Korea
Tel: + 82 2 958 5455
Fax: + 82 2 958 5449
Email: dwkum@kistmail.kist.re.kr
http: //www.ifsm.umn.edu/members.html#

- **Mexican Society for Electron Microscopy**

Dr Raul Herrera Becerra (President)
Av. Universidad No. 1953
Edificio 17 Depto. 102
Copilco El Bajo
Delegación Coyoacán
04350 México, D. F.
Tel: +52 614 439 1172
Fax: +52 614 481 0812
Email: rherrera@fenix.ifisicacu.unam.mx or raulhb@servidor.unam.mx

- **Microscopy New Zealand Inc**

Allan Mitchell (President)
Otago Centre for Electron Microscopy
School of Medical Sciences
PO Box 913
Dunedin
New Zealand
Email: mailto: allan.mitchell@stonebow.otago.ac.nz
Tel: 0064 3 479 5642

- **Joy Woods (Vice president)**

Email: woods@wronz.org.nz
Tel: 0064 3 325 2421

- **Peter Smith (Secretary)**

Email: mailto: smithp@agresearch.cri.nz
Tel: +64 4 528 1380

- **Karen Reader (Treasurer)**
  Email: karen.reader@agresearch.co.nz
  Tel: 0064 4 528 6089
- **Reproduction Group**
  AgResearch Wallaceville
  Ward Street
  Upper Hutt
  New Zealand
  http: //microscopynz.otago.ac.nz/
- **Polish Commission for Electron Microscopy**
  Prof Wieslawa Biczysko, MD PhD (President)
  Electron Microscopy Laboratory
  Department of Clinical Pathomorphology
  Karol Marcinkowski University of Medical Sciences
  Przybyszewskiego 49
  60-355 Poznan
  Poland
  Tel: +48 61 8691483
  Fax: +48 61 8691509
- **Portuguese Society for Electron Microscopy and Cell Biology**
  Maria de Lourdes Pereira (President)
  Maria Conceição Santos (Secretary)
  Violante Gomes (Treasurer)
  SecretarySPME-BC for2004
  Departamento de Biologia
  Universidade de Aveiro
  Campus Universitário de Santiago
  3810-193 Aveiro
  Tel.: 234 370780
  Fax: 234 426408
  http: //www.spme-bc.pt
  Email: lpereira@bio.ua.pt

- **Russian Society for Electron Microscopy**
  Russian Academy of Sciences
  Professor NA Kiselev
  Leninski Pr 59
  117333 Moscow Russia
  Tel: +7 095 1351520
  Fax: +7 095 1351011
  Email: roddatis@ns.crys.ras.ru
- **SCANDEM Nordic Microscopy Society**
  Eva Olsson (President)
  Experimental Physics, Microscopy and Microanalysis
  Chalmers University of Technology
  SE-41296 Göteborg
  Sweden
  Tel: +46 31 772 3247
  Fax: +46 31 772 3224
  Email: eva.olsson@fy.chalmers.se
  http: //www.scandem.org/
  Vice President & Secretary General:
  Anette Gunnaes
  Department of Physics
  Centre for Materials Science
  University of Oslo
  Gaustadalleen 21
  NO-0349 Oslo
  Norway
  Tel: +47 2284 0696
  Fax: +47 2284 0651
  Email: mailto: leonora@fys.uio.no
  Treasurer:
  Salla Marttila
  Department of Crop Science

Swedish University of Agricultural Sciences
SE-23053 Alnarp
Sweden
Tel: +46 40415522
Fax: +46 40415519
Email: salla.marttila@vv.slu.se

- **Microscopy Society (Singapore)**

Dr Mark Yeadon (President)
Institute of Materials Research and Engineering
3 Research Link
Singapore 117602
Fax: +65 6872 0785
Email: m-yeadon~@imre.org.sg
http: //www.microscopy.org.sg/
Associate Professor Mary ML Ng
Electron Microscopy Unit (MS5, Level 1)
Faculty of Medicine
National University of Singapore
10 Kent Ridge Crescent
SINGAPORE 119260
FAX: +65 776 6872
Email: micngml@nus.edu.sg

- **Slovenian Society for Microscopy**

President Dr Roc Romih
Institute of Cell Biology, Faculty of Medicine,
University of Ljubljana,
Lipiceva 2,
SI-1000 Ljubljana,
Slovenia.
Tel: +386 1 543-76-87
Email: rok.romih@mf.uni-lj.si
Vice President Dr. Miran Ceh.

Katarina Merzel
Treasurer SSM
Institute of Biophysics, Faculty of Medicine,
University of Ljubljana
Lipiceva 2
SI-1000 Ljubljana
Slovenia
Tel: +386 1 543-76-00,
Fax: +386 1 431-51-27
E-mail: mailto: %20katarina.merzel@biofiz.mf.uni-lj.si

- **Microscopy Society of Southern Africa**
  Mr Alan Hall
  Laboratory for Microscopy and Microanalysis
  NW11 Building
  University of Pretoria
  PRETORIA 0002
  South Africa
  Tel: +27 12 420 3896/2075
  Fax: +27 12 362 5150
  Email: AHall@nsnper1.up.ac.za
  http: //www.ifsm.umn.edu/members.html#
- **Spanish Microscopy Society**
  (Sociedad de Microscopia de Espana (SEME))
  J.M. González Calbet,
  Dpto. Química Inorgánica,
  Facultad de Ciencias Químicas,
  Ciudad Universitaria, s/n,
  Universidad Complutense,
  E-28040 Madrid
  Spain.
  Tel: (+34)-91-394.43.42
  Fax: (+34)-91-394.43.52

Email: jgcalbet@quim.ucm.es

http: //www.icmm.csic.es/esisna/sme/sme.htm

- **Swiss Society for Optics and Microscopy**

  Dr Kurt Pulfer

  President SSOM

  Solvias AG

  Klybeckstrasse 191

  WKL-127.6.34

  Postfach CH-4002 Basel, Switzerland

  Tel: +41 1 686 6221

  Fax: +41 1 686 6096

  Email: kurt.pulfer@solvias.com

  Treasurer:

  Dr Marcel Duggelin

  Universitat Basel

  Zentrum fur Mikroskopie

  Pharmazentrum

  Klingelbergstrasse 50

  CH-4056 Basel

  Switzerland

  Tel: 061-267-1402

  Fax: 061-267-1410

  Email: mailto: %20marcel.dueggelin@unibas.ch

  http: //www.ifsm.umn.edu/members.html#

- **Electron Microscopy Society, Taipei, China**

  Prof. Ji-Jung Kai, President and Chairman

  Department of Engineering and System Science

  National Tsing Hua University

  Hsinchu, Taiwan 300

  ROC

  Tel: +886 3 5742662

  Fax: +886 3 5734066

Email: mailto: %20jjkai@ess.nthu.edu.tw
http: //communities.msn.com/microscopytw
Prof. Shiang-Ming Yu (Vice President)
Institute of Anatomy and Cell Biology
National Yang-Ming University
Taipei, Taiwan 112, R.O.C.
Prof. Li Chang (General Secretary)
Department of Materials Science and Engineering
National Chao-Tong University
Hsinchu, Taiwan 300, R.O.C.

- **Electron Microscopy Society of Thailand**
  Professor Waykin Nopanitaya
  c/- STREC, Institute Building 2
  Chulalongkorn Soi 62
  Phayathai Road
  Bangkok 10330
  Thailand
  Tel: +66 218 8030
  Fax: +66 218 6770
  Assoc. Prof Dr Pramote Vanittanakom (President)
  Department of Pathology
  Faculty of Medicine
  Chiang Mai University
  Chiang Mai 50200
  Thailand
  Tel: +66 53 945442
  Fax: +66 53 217144
  Email: pvanitta@mail.med.cmu.ac.th
- **Turkish Society for Electron Microscopy**
  Prof Dr Suzan Daglioglu (President)
  Prof Dr Serap Arbak (General Secretary)
  Dr Feriha Ercan

Ordinary Members:
Prof Dr Sehnaz Bolkent
Prof Dr Tangul San
University of Istanbul
Faculty of Veternary Medicine
Department of Histology & Embryology
Avcilar
Istanbul
Turkey
Email: suzandag@istanbul.edu.tr
http: //www.temd.org/

- **Venezuelan Society for EM**

Dra Caribay Urbina—President
Departamento de Quimica
Escuela de Quimica
Universidad Central de Venezuela
Direccion electronica:
E-mail curbina@electra.ciens.ucv.ve
Sociedad Venezolana de Microscopia Electronica
Sede Secretarial
Facultad de Ciencias
Universidad Central de Venezuela
AP 47140 Caracas 1041-A
Venezuela
Tel: +58 2 6051262/6051606
Fax: +58 2 6930694
E-mail svme@electra.ciens.ucv.ve

4
Global Nanotech Labs and Research Facilities

SPM and Nanomanipulation

The Scanning Tunelling Microscope (STM) was invented by Binnig and Rohrer at the IBM Zürich laboratory in the early 1980s, and won them a Nobel Prize four years later. The principles of the instrument can be summarized with the help. A sharp conducting probe, typically made from tungsten is placed very close to a sample, that must also be a conductor. The tip is biased with respect to the sample as shown in the figure. The tip can be moved towards or away from the sample, i.e., in the *-z* or *+z* directions in the figure, by means of piezoelectric actuators. At Ångstrom-scale distances, a quantum-mechanical effect, called *tunneling*, causes electrons to flow across the tip/ sample gap, and a current can be detected. To first approximation, the tunneling current depends exponentially on the distance between tip and sample. This current is kept constant by a feedback circuit that controls the piezoelectric actuators. Because of the current/distance relationship, the distance is also kept constant, and the z accuracy is very high because any small z variation causes an exponential error in the current. Additional piezo motors drive the tip in a xy scanning motion. Since the tip/ sample gap is kept constant by the feedback, the scanning tip traverses a surface parallel to the sample surface. The result of a scan is a $z(x, y)$ terrain map, with enough resolution to detect atomic-scale features of the sample, as indicated diagrammatically in the figure. Various instruments analogous to the STM have been built. They exploit physical properties other than the tunneling effect on which the STM is based. The most common of these other instruments is the Atomic Force Microscope (AFM), which is based on interatomic forces. All of these instruments are collectively known as Scanning Probe Microscopes (SPMs). The principles of operation of the AFM.. The forces between atoms in the tip and sample cause a deflection of the cantilever that carries the tip. The amount of deflection is measured by means of a laser beam bouncing off the top of the cantilever. (There are other schemes for measuring deflection.) The force depends on the tip/sample gap, and therefore servoing on the force ensures that the distance is kept constant while scanning, as in the STM. The AFM does not require conducting tips and samples, and therefore has wider applicability than the STM. If the tip of the AFM is brought very close to the sample, at distances of a few Å, repulsive forces prevent the tip from penetrating the sample. This mode of operation is called contact mode. It provides good resolution but cannot be used with delicate samples, *e.g.* biomaterials, which are damaged by the contact forces. Alternatively, the AFM tip can be placed at distances in the order of several nm or tens of nm, where the interatomic forces between tip and sample are attractive. The tip is vibrated at a frequency near the resonance frequency of the cantilever, in the kHz range. The tip/sample force is equivalent to a change in the spring constant of the cantilever, and causes a change in its resonance frequency. This change can be used as the error signal in the feedback circuit that controls the tip. (There are alternative detection schemes.) This mode of operation is called non-contact mode. It has poorer resolution than contact mode but can be used with delicate samples. Although the SPM is just about twenty years old, it has had a large scientific impact.

Since the early days of the SPM it was known that tip/sample interaction could produce changes in both tip and sample. Often these were undesirable, for example, a blunt probe due to a crash into the sample. But it soon became clear that one could produce new and desirable features on a sample by using the tip in a suitable manner. One of the first demonstrations was done by Becker and co-workers at Bell Labs, who managed to create nanometer-scale germanium structures on a germanium surface by raising the voltage bias of an STM tip. Much of the subsequent work falls under the category of nanolithography and will not be discussed here. In the following subsections we survey nanomanipulation research involving the SPM.

Pushing and pulling operations are not widely used in macrorobotics, although there has been interesting work on orienting parts by pushing, done by Matt Mason at CMU, Ken Goldberg at USC, and others. The techniques seem suitable for constructing 2-D structures. Interatomic attractive forces were used by Eigler *et al.* at IBM Almadén to precisely position xenon atoms on nickel, iron atoms on copper, platinum atoms on platinum, and carbon monoxide molecules on platinum. The atoms are moved much like one displaces a small metalic object on a table by moving a magnet under the table. The STM tip is placed sufficiently close to an atom for the attractive force to be larger than the resistance to lateral movement. The atom is then pulled along the trajectory of the tip. Eigler's experiments were done in ultra high vacuum (UHV) at very low temperature (4K). Low temperature seems essential for stable operation. Thermal noise destroys the generated patterns at higher temperatures. Lateral repulsive forces were used by Güntherodt's group at the University of Basel to push fullerene (C_{60}) islands of ~ 50nm size on flat terraces of a sodium chloride surface, in UHV, at room temperature, with a modified AFM. The ability to move the islands depends strongly on species/substrate interaction, e.g., C_{60} does not move on gold, and motion on graphite destroys the islands. Lateral forces opposing the motion are analogous to friction in the macroworld, but cannot be modeled simply by Coulomb friction or similar approaches that are used in macrorobotics. Mo at IBM Yorktown rotated pairs of antimonium atoms between two stable orientations 90 degrees apart. This was done in UHV at room temperature, on a silicon substrate, by scanning with an STM tip with a higher voltage than required for imaging. The rotation was reversible, although several scans were sometimes necessary to induce the desired motion. Samuelson's group at the University of Lund succeeded in pushing galium arsenide (GaAs) nanoparticles of sizes in the order of 30 nm on a GaAs substrate at room temperature in air. The sample is first imaged in non-contact AFM mode. Then the tip is brought close to a nanoparticle, the feedback is turned off and the tip is moved against the nanoparticle. Schaefer *et al.* at Purdue University push gold clusters with an AFM in a nitrogen environment at room temperature. They first image the clusters in non-contact mode, then remove the tip oscillation voltage, and sweep the tip across the particle in contact with the surface and with the feedback disabled.

A similar technique is being used at USC's Laboratory for molecular robotics to push colloidal gold nanoparticles with 15 nm diameters on a mica substrate at room temperature and in ambient air. We image in non-contact mode, then disable the feedback and push by moving the tip in a single line scan, without removing the tip oscillation voltage. The right a "USC" pattern written with gold nanoparticles. The *z* coordinate is encoded as brightness in this figure. On the left is the original random pattern, before manipulation with the AFM. Smaller objects have been arranged into prescribed patterns at room temperature by Gimzewski's group at IBM's Zürich laboratory. They push molecules at room temperature in UHV by using an STM. They have succeeded in pushing porphyrin molecules on copper, and more recently they have arranged bucky balls (i.e., C_{60}) in a linear pattern, using an atomic step in the copper substrate as a guide. C_{60} molecules on silicon also have been pushed with an STM in UHV at room temperature by Maruno *et al.* in Japan, and Beton *et al.* in the U.K. In Maruno's approach the STM tip is brought closer to the surface than in normal imaging mode, and then scan across a rectangular region with the feedback turned off. This causes many probe crashes. In Beton's approach the tip also is brought close to the surface, but the sweep is done with the feedback on and a high value for the tunneling current. Their success rate is in the order of only 1 in 10 trials.

Much of industrial macrorobotics is concerned with pick and place operations, which typically do not require very precise positioning or fine control. There are a few examples of experiments in which atoms or molecules are transferred to SPM tips, these are moved, and the atoms transferred back to the surfaces. Eigler *et al.* succeeded in transferring xenon atoms from platinum or nickel surfaces to an STM tip by moving the tip sufficiently close for the adsorption barriers of surface and tip to be comparable. An atom may leave the surface and become adsorbed to the tip, or vice-versa. Benzene molecules also have been transferred to and from tips. Eigler's group also has been able to transfer xenon atoms between an STM tip and a nickel surface by applying voltage pulses to the tip. This is attributed to electromigration, caused by the electric current flowing in the tunneling junction. All of Eigler's work has been done in UHV at 4K. Avouris' group, at IBM Yorktown, and the Aono group in Japan have transferred silicon atoms between a tungsten tip and a silicon surface in UHV at room temperature, by applying voltage pulses to the tip. The mechanism for the transfer is believed to be field-induced evaporation, perhaps aided by chemical phenomena at the tip/surface interface in Avouris work.

This is the most sophisticated form of macrorobotic motion. It involves fine motions, as in a peg-in-hole assembly, in which there is accomodation and often force control, for example to ensure contact between two surfaces as they slide past each other. Compliance is crucial for successful assembly operations in the presence of spatial and other uncertainties, which are unavoidable. The study of the nanoscale analog of compliant motion seems to be virgin territory. We speculate that the analog of compliance is chemical affinity between atoms and molecules. It is suspected that such "chemical compliance" may prove essential for nanoassembly operations at room temperature, in the presence of thermal noise. It seems likely that successful assembly of nanoscale components will require a combination of precise positioning and chemical compliance. Therefore, work on self-assembling structures is relevant.

To first approximation, an SPM is a 3 degree-of-freedom robot. It can move in x, y, and z, but cannot orient its tip, which is the analog of a macrorobotic hand. (How could a 6 degree-of-freedom SPM be built, and what could be done with it, are interesting issues.) The vertical displacement is controlled very accurately by a feedback loop involving tunneling current or force (or other quantities for less-common SPMs). But nanoscale x, y motion over small regions (e.g., within a 5 micron square) is primarily open-loop, because of a lack of suitable sensors that can be used in a feedback scheme. Accurate horizontal motion relies on calibration of the piezoelectric actuators, which are known to suffer from a variety of problems such as creep and hysteresis. In addition, thermal drift of the instrument is very significant. At room temperature a drift of one atomic diameter per second is common, which means that manipulation of atomic objects on a surface is not unlike picking parts from a conveyor belt. Thermal drift is negligible if the SPM is operated at very low temperatures, and all the experiments in atomic-precision manipulation to date have been done at 4K. This involves complex technology and is clearly undesirable. For room-temperature nanomanipulation, drift, creep and hysteresis must be taken into account. Ideally, compensation should be automatic. Research is under way at USC on how to move accurately an SPM tip in the presence of all these sources of error. To complicate matters further, the SPM often must operate in a liquid environment, especially if the objects to be manipulated are biological. Little is known about nanomanipulation in liquids.

The SPM functions both as a manipulator and a sensing device. The lack of a direct and independent means of establishing "ground truth" while navigating the tip causes delicate problems. The SPM commands are issued in instrument or robot coordinates, and sensory data also are acquired in robot coordinates, whereas manipulation tasks are expressed in sample or task coordinates. These two coordinate systems do not coincide, and indeed are in relative motion due to drift. Accurate motion in task coordinates may be achieved by a form of visual servoing, by tracking features of the sample and moving relative to them. We move the tip to the approximate position of a nanoparticle to be pushed, and then search for it through single line scans. Feature tracking assumes that features are stationary in task coordinates. This assumption may fail at room temperature if the features are sufficiently small, because of thermal agitation. Hence, atomic manipulation at room temperature may require artificially-introduced features that can be tracked to establish a task coordinate system. It must also deal with the spatial uncertainty associated with the thermal motion of the atoms to be moved. Larger objects such as molecules and clusters have lower spatial uncertainty, and should be easier to handle.

The SPM output signal depends not only on the topography of the sample but also on the shape of the tip, and on other characteristics of the tip/sample interaction. For example, the tunneling current in an STM depends on the electronic wavefunctions of sample and tip. To first approximation one may assume that the tip and sample are in contact. Under this assumption one can use configuration-space techniques to study the motion of the tip]. The procedure in 2-D. On the top of the figure we consider a tip with a triangular end and a square protrusion in the sample. In the bottom we consider a tip with a semi-circular end and the same protrusion. We choose as reference point for the tip its apex, and reflect the tip about the reference point. The configuration-space obstacle that corresponds to the real-space protrusion obstacle is obtained by sweeping the inverted tip over the protrusion so that the apex remains inside the obstacle. Mathematically, we are calculating the Minkowski sum of the inverted tip and the protrusion. The path of the tip in its motion in contact with the obstacle is the detected topographical signal. As shown in the figure, the sensed topography has been broadened by the dimensions of the tip. Note, however, that the detected height is correct. (Minkowski operations and related mathematical morphology tools were introduced in the SPM literature only recently.) Tip effects are sometimes called "convolution", by analogy with the broadening of an impulse passing through a linear system, and one talks of "deconvolving" the image to remove tip effects. The configuration-space analysis outlined above is purely geometric and provides only a coarse approximation to the SPM output signal. More precise calculations may be performed numerically. For example, in contact AFM we can assume specific atomic distributions for the tip and sample, and a specific form for the interatomic forces, and compute the resulting tip/sample force. A major issue in tip-effect compensation is that the shape of the probe is not known, and indeed may vary during operation. For example, atoms may be adsorbed on the tip or lost because of the contact with the sample. The most promising approach for dealing with tip effects consists of estimating the tip shape by using it to image known features. If necessary, artificial features may be introduced into the scene for tip estimation purposes. The estimated tip shape can then be used to remove (at least in part) the tip effects from the image by using Minkowski operations. Removal procedures that take into account more sophisticated, non-geometric effects do not appear to be known.

Sensor fusion techniques may be used, at least in principle, for increasing the quality of the sensory data, because it is possible to access several signals during an SPM scan. For example, vertical and lateral force (akin to friction) can be recorded simultaneously in typical AFMs. To our knowledge, sensor fusion has not been attempted in SPM technology. It is clear that faithful sensory data should facilitate manipulation tasks. What is not clear is whether clever manipulation strategies can compensate for the imperfections of SPM data.

The SPM tip is the primary end effector in nanomanipulation. A plain, sharp tip seems to be adequate for most pushing operations. In some cases it may also suffice for picking and depositing objects, especially in conjunction with electrostatic forces generated by applying a suitable bias to the tip. This requires both a conducting tip and a conducting substrate. Pick-and-place tasks, however, usually require the nanoscale analog of a gripper. Very little is known about molecular grippers. One can think of a nanogripper as a molecule or cluster of molecules that are attached to a tip and are capable of picking up and transporting other molecules or particles. (Tips with attached molecules are said to be functionalized.) Ideally these grippers should be switchable, so as to pick and release objects on command. Candidates for grippers are certain molecules such as cyclodextrins, which have large cavities that can carry other molecules. Coating a tip with strands of DNA may also permit picking up objects that are coated with complementary strands. In both of these examples, switching the gripper on and off is not a solved problem. Techniques for changing SPM tips automatically do not exist. Changes must be done manually, and it is very difficult or impossible to maintain sample registration, *i.e.*, to return to the same position on the sample after tip replacement. This implies that one often must image using a tip with an attached gripper. Again, little is known about imaging with such functionalized tips.

In the macroworld fixtures are often necessary to hold mechanical parts during assembly and other manufacturing operations. In micromechanics sacrificial layers are used as scaffolding to fabricate certain microelectromechanical systems (MEMS). At the nanoscale, the analogs of fixtures are substrates that ensure that certain objects remain fixed during manipulation, while others are allowed to move. Substrate selection seems to be highly dependent on the chemistry of the objects being manipulated.

In macrorobotics the physical processes involved are mechanical and relatively well understood. At the nanoscale, the processes are chemical and physical, and still an area of active research. In addition, nanomanipulation takes place in several different environments, such as liquids, air or UHV. The environment has a strong influence on the physics and chemistry of the processes. Nanomanipulation is not restricted to mechanical interactions. For example, light, electrostatic fields, and the pH of a liquid all are candidates for controlled interaction with nanoparticles.

High level programing and planning systems are highly desirable, and indeed essential for assembling complex structures. One must begin with relatively low-level programing primitives and build upon them constructs at a higher level of abstraction. High-level commands must be compiled into low-level primitives. This compilation may involve sophisticated computations, for example to ensure collision-free paths in an environment with large spatial uncertainty. What are the relevant high-level manipulation tasks? For example, what is the nanoscale equivalent of a peg-in-hole insertion? In short, we may need to adapt much of what is known about macrorobotics to the nanoworld. It is likely that new concepts will also be needed, because the physics and chemistry of the phenomena and objects are quite different in these two worlds.

What hardware primitives are suitable as building blocks? The nanotechnology literature suggests hardware primitives based on DNA structures such as those built in Seeman's lab, proteins, and diamondoid structures. Biomaterials such as DNA and proteins may be too flimsy, whereas diamondoid structures are expected to be very strong. No experiments have yet been reported in which any of these components are successfully assembled into a composite structure. Which tasks should one attempt first? What should one try to build? Here the options are in the realms of electronics, photonics, mechanics, or biomaterials. On-going research at USC is attempting to build nanowires as assemblies of gold particles, and arrays of nanostructures with photonic properties on patterned semiconductor substrates.

Nanorobotics manipulation with SPMs is a promising field that can lead to revolutionary new science and technology. But it is clearly in its infancy. Typical nanomanipulation experiments reported in the literature involve a team of very skilled, Ph.D.-level researchers working for many hours in a tightly-controlled environment (typically in ultra high vacuum and at low temperature, often 4K) to build a pattern with tens of nanoparticles. It still takes the best groups in the world some 10 hours to assemble a structure with about 50 atoms. This is simply too long-changes will occur in many systems, *e.g.* contamination or oxidation of the components, on a timescale that will constrain the maximum time available for nanomanipulation. Requiring all operations to take place at 4K and in UHV also is not practical for widespread use of nanomanipulation. In short, nanomanipulation today is more of an experimental tour-de-force than a technique that can be routinely used. It is clear that complex tasks cannot be accomplished unless the SPM is commanded at a higher level of abstraction. Compensation for instrument inaccuracies should be automatic, and the user should be relieved from many low-level details. Building a high level programing system for nanomanipulation is a daunting task. The various component technologies needed for nanomanipulation must be developed and integrated. These technologies include:

substrates that serve as nanofixtures or nanoworkbenches on which to place the objects to be manipulated; tips, probes and molecules that serve as grippers or end-effectors; chemical and physical nanoassembly processes; primitive nanoassembly operations that play a role analogous to macroassembly tasks such as peg-in-hole insertion; methods for exploiting self-assembly to combat spatial uncertainty, in a role analogous to mechanical compliance in the macroworld; suitable hardware primitives for building nanostructures; and algorithms and software for sensory interpretation, motion planning, and driving the SPM. This is a tall order, and requires an interdisciplinary approach that combines synergistically the knowledge and talents of roboticists and computer scientists with those of physicists, chemists, materials scientists, and perhaps biologists. SPM-based assembly methods face a major scale-up challenge. Building complex structures one atom (or even one nanoparticle) at a time is very time consuming. We believe that SPMs will have applications in the exploration of new structures, which may later be mass produced by other means. This is the nanoworld analog of rapid protyping technologies such as stereolithography that are becoming popular at the macroscale.

There are at least two approaches for fighting the serial nature of SPM manipulation. The first involves the use of large arrays of SPMs on a chip. These chips are being developed at Cornell and Stanford. Programing such arrays for coordinated assembly tasks poses interesting problems. The second approach is subtler, and consists of using the SPM to construct structures that are capable of self-replication. The best known such structures involve DNA, but other systems also exist. Self-replication is inherently an exponential process. In summary, nanomanipulation with SPMs may have a revolutionary impact on science, technology, and the way we live. To fully exploit its potential we will have to develop powerful systems for programing nanorobotic tasks. Much of what is known in macrorobotics is likely to be relevant, but may have to be adapted to the nanoworld, where phenomena and structures are quite different from their macroscopic counterparts. Research at USC and elsewhere is progressing, with promising results. Nanomanipulation, perhaps coupled with self-assembly, is expected to eventually succeed in building true nanorobots, *i.e.*, devices with overal dimensions in the nanometer range and capable of sensing, "thinking", and acting. Complex tasks are likely to require a group of nanorobots working cooperatively. This raises interesting issues of control, communications, and programing of robot "societies".

Overview of Scanning Probe Microscopy Techniques

Scanning Probe Microscopy has enabled researchers to image surfaces at the nanometer scale. Rather than using a beam of light or electrons, SPM uses a fine probe that is scanned over a surface (or the surface is scanned under the probe). By using such a probe, researchers are no longer restrained by the wavelength of light or electrons. The resolution obtainable with this technique can resolve atoms, and true 3-D maps of surfaces are possible. Scanning Probe Microscopy is a general term, used to describe a growing number of techniques that use a sharp probe to scan over a surface and measure some property of that surface. Some examples are STM (scanning tunneling microscopy), AFM (atomic force microscopy), and NSOM (Near-Field Scanning Optical Microscopy).

Scanning Tunneling Microscopy

The development of the family of scanning probe microscopes starts with the original invention of the STM in 1981. Gerd Binnig and Heinrich Rohrer developed the first working STM while working at IBM Zurich Research Laboratories in Switzerland. This instrument would later win Binnig and Rohrer the Nobel prize in physics in 1986.

The STM works by scanning a very sharp metal wire tip over a surface. By bringing the tip very close to the surface, and by applying an electrical voltage to the tip or sample, we can image the surface at an extremely small scale – down to resolving individual atoms.

The STM is based on several principles. One is the quantum mechanical effect of tunneling. It is this effect that allows us to "see" the surface. Another principle is the piezoelectric effect. It is this effect that allows us to precisely scan the tip with angstrom-level control. Lastly, a feedback loop is required, which monitors the tunneling current and coordinates the current and the positioning of the tip.

Tunneling

Tunneling is a quantum mechanical effect. A tunneling current occurs when electrons move through a barrier that they classically shouldn't be able to move though. In classical terms, if you don't have enough energy to move "over" a barrier, you won't. However, in the quantum mechanical world, electrons have wavelike properties. These waves don't end abruptly at a wall or barrier, but taper off quite quickly. If the barrier is thin enough, the probability function may extend into the next region, though the barrier! Because of the small probability of an electron being on the other side of the barrier, given enough electrons, some will indeed move through and appear on the other side. When an electron moves though the barrier in this fashion, it is called tunneling. Quantum mechanics tells us that electrons have both wave and particle like properties. Tunneling is an effect of the wavelike nature. The top image shows us that when an electron (the wave) hits a barrier, the wave doesn't abruptly end, but tapers off very quickly—exponentially. For a thick barrier, the wave doesn't get past. The bottom image shows the senario if the barrier is quite thin (about a nanometer). Part of the wave does get through, and therefore some electrons may appear on the other side of the barrier..

Because of the sharp decay of the probability function through the barrier, the number of electrons that will actually do this is very dependent upon the thickness of the barrier. The actual current through the barrier drops off exponentially with the barrier thickness.

To extend this description to the STM: The starting point of the electron is either the tip or sample (depending on the setup of the instrument). The barrier is the gap (air, vacuum, liquid), and the second region is the "other side" – tip or sample, again, depending on the experimental setup. By monitoring the current through the gap, we have very good control of the tip-sample distance.

Piezoelectric Materials

The piezoelectric effect was discovered by Pierre Curie in 1880. The effect is created by squeezing the sides of certain crystals, such as quartz or barium titanate. The result is the creation of opposite charges on the sides. The effect can be reversed as well; by applying a voltage across a piezoelectric crystal, it will elongate or compress.

These materials are used to scan the tip in an STM, and most other scanning probe techniques. A typical piezoelectric material used in STMs is PZT (Lead Zirconium Titanate).

Electronics and the Feedback Loop

Obviously, you need electronics to measure the current, scan the tip, and translate this information into a form that we can use. A feedback loop constantly monitors the tunneling current and makes adjustments to the tip to maintain a constant tunneling current. These adjustments are recorded by the computer and presented as an image in the STM software. Such an setup is called a "constant current" image. In addition, for very flat surfaces, the feedback loop can be turned off and only the current is displayed. This is a "constant height" image.

The Result

The STM is cabable of acquiring remarkable images on the most extreme scale, easily resolving atomic structure in the right environments. For some very interesting work with STM, see the IBM image gallery.

The Quantum Corral

This STM image shows the direct of standing-wave patterns in the local density of states of the Cu(111) surface. These spatial oscillations are quantum-mechanical interference patterns caused by scattering of the two-dimensional electron gas off the Fe adatoms and point defects.

Atomic Force Microscopy

The Atomic Force Microscope was developed to overcome a basic drawback with STM—that it can only image conducting or semiconducting surfaces. The AFM, however, has the advantage of imaging almost any type of surface, including polymers, ceramics, composites, glass, and biological samples. Binnig, Quate, and Gerber invented the Atomic Force MIcroscope in 1985. Their original AFM consisted of a diamond shard attached to a strip of gold foil. The diamond tip contacted the surface directly, with the interatomic van der Waals forces providing the interaction mechanism. Detection of the cantilever's vertical movement was done with a second tip—an STM placed above the cantilever.

AFM probe deflection

Today, most AFMs use a laser beam deflection system, introduced by Meyer and Amer, where a laser is reflected from the back of the reflective AFM lever and onto a position-sensitive detector. AFM tips and cantilevers are microfabricated from Si or Si_3N_4. Typical tip radius is from a few to 10s of nm.

Beam deflection system, using a laser and photodector to measure the beam position.

Measuring forces

Because the AFM relies on the forces between the tip and sample, knowing these forces is important for proper imaging. The force is not measured directly, but calculated by measuring the deflection of the lever, and knowing the stiffness of the cantilever. Hook's law gives F = -kz, where F is the force, k is the stiffness of the lever, and z is the distance the lever is bent.

AFM Modes of operation

Because of AFM's versatility, it has been applied to a large number of research topics. The Atomic Force Microscope has also gone through many modifications for specific application requirements.

Contact Mode

The first and foremost mode of operation, contact mode is widely used. As the tip is raster-scanned across the surface, it is deflected as it moves over the surface corrugation. In constant force mode, the tip is constantly adjusted to maintain a constant deflection, and therefore constant height above the surface. It is this adjustment that is displayed as data. However, the ability to track the surface in this manner is limited by the feedback circuit. Sometimes the tip is allowed to scan without this adjustment, and one measures only the deflection. This is useful for small, high-speed atomic resolution scans, and is known as variable-deflection mode.

Because the tip is in hard contact with the surface, the stiffness of the lever needs to be less that the effective spring constant holding atoms together, which is on the order of 1-10 nN/nm. Most contact mode levers have a spring constant of < 1N/m.

Lateral Force Microscopy

LFM measures frictional forces on a surface. By measuring the "twist" of the cantilever, rather than merely its deflection, one can qualitatively determine areas of higher and lower friction.

Noncontact mode

Noncontact mode belongs to a family of AC modes, which refers to the use of an oscillating cantilever. A stiff cantilever is oscillated in the attractive regime, meaning that the tip is quite close to the sample, but not touching it (hence, "noncontact"). The forces between the tip and sample are quite low, on the order of pN (10^{-12} N). The detection scheme is based on measuring changes to the resonant frequency or amplitude of the cantilever.

Dynamic Force / Intermittant-contact / "tapping mode" AFM

Commonly referred to as "tapping mode" it is also referred to as intermittent-contact or the more general term Dynamic Force Mode (DFM).

A stiff cantilever is oscillated closer to the sample than in noncontact mode. Part of the oscillation extends into the repulsive regime, so the tip intermittently touches or "taps" the surface. Very stiff cantilevers are typically used, as tips can get "stuck" in the water contamination layer.

The advantage of tapping the surface is improved lateral resolution on soft samples. Lateral forces such as drag, common in contact mode, are virtually eliminated. For poorly adsorbed specimens on a substrate surface the advantage is clearly seen.

Force Modulation

Force modulation refers to a method used to probe properties of materials through sample/tip interactions. The tip (or sample) is oscillated at a high frequency and pushed into the repulsive regime. The slope of the force-distance curve is measured which is correlated to the sample's elasticity. The data can be acquired along with topography, which allows comparison of both height and material properties.

Phase Imaging

In Phase mode imaging, the phase shift of the oscillating cantilever relative to the driving signal is measured. This phase shift can be correlated with specific material properties that effect the tip/sample interaction. The phase shift can be used to differentiate areas on a sample with such differing properties as friction, adhesion, and viscoelasticity. The techniques is used simultaneously with DFM mode, so topography can be measured as well.

Near-field Scanning Optical Microscopy (NSOM)

SNOM offers the use of a very small light source as the imaging mechanism. By using a quasipoint light source with a diameter much smaller than the wavelength of light, one can achieve resolutions better than the diffraction limit. The probe, however, must be very close to the surface; much closer than the wavelength of the light. This region is the "Near-Field" and hence the name of the technique.

Typically, laser light is fed to the aperture via an optical fiber. The aperture can be a tapered fiber coated with a metal (such as Al), a microfabricated hollow AFM probe, or a tapered pipette. Normally, the size of the point light source determines the resolution obtainable.

There are two types of feedback typically used to maintain the proper working distance of the probe to the sample. One method is quite similar to how feedback works with an AFM—by using a cantilevered probe, the normal force is monitored, typically by using a beam-deflection setup as in most AFMs. The second method uses a tuning fork. By attaching the fiber to a tuning fork, which oscillates at its resonant frequency, can can monitor changes in the amplitude as the tip moves over the surface. The tip is moved laterally, and this techniques is normally referred to as "shear-force" feedback.

Depending upon the sample being imaged, there are multiple modes of operation for NSOM.

Transmission: Lightsource travels through the probe aperture, and transmits through sample. Requires a transparent sample.

Reflection: Lightsource travels through the probe aperture, and reflects from the surface. Lower light intensity, and tip-dependent, but allows for opaque samples.

Collection: Sample is illuminated from large outside light source, and the probe collects the reflected light.

Illumination / Collection: The probe both illuminates the sample and collects the reflected light.

Detection of the signal can be handled a number of different ways: Spectrometer, APD (Avalanche Photo Diode), Photomultiplier Tube, or CCD

Data Display: several contrast mechanisms in NSOM can be used, including polarization, topography, birefringence, index of refraction, fluorescence, wavelength dependence, and reflectivity.

An example of a commerical NSOM is the WITec AlphaSNOM.

The Nanosurf® EasyScan 2 AFM and STM

The Smart, Modular and Robust Scanning Probe Microscopes

Configure your own AFM or STM that suits your needs at an affordable cost.

The highly modular easyScan 2 line allows you to move from a very affordable STM system to a highly functional multiple mode AFM/STM package. You can configure what you need, or you can add options later.

The Swiss-made system is manufactured near the birthplace of the STM. High quality, precision machining, and a smart easy-to-use design makes the easyScan 2 the perfect choice for teaching or research.

The easyScan 2 system is built around the elegant easyScan 2 controller. The system and software can be configured to control either of 2 STM microscope heads, or any of 3 different AFM heads. Auxiliary boards and functional upgrades are available to expand the system's capabilities.

The easyScan 2 STM:

The simple elegance and affordability of the STM has been experienced by hundreds of researchers and educators around the world. The easyScan 2 offers new capabilities not available with the first model, including many upgrade options.

The easyScan 2 AFM:

One can easily upgrade the STM to the AFM, or simply bypass the STM option altogether. The new easyScan 2 AFM lets you get started with atomic force microscopy at a low cost, and you have the ability to upgrade functions later one.

Advanced easyScan 2 options:

The easyScan 2 AFM can be upgraded by adding dynamic force functions, dual-view video, sample translation, a signal access monitor, software capabilities and many other options.

Nonopolis Indusrial Partners

SPECTRO will launch its next generation, high performance CIROS VISION ICP-OES CCD optical emission spectrometer at Pittcon 2004, 8-11 March in Chicago, USA. The new SPECTRO CIROS VISION ICP system is the most powerful, flexible and fastest ICP spectrometer in the world today and is the only system with a broad emission range to enable complete chemical elemental analysis. New highlight features and functionality include a significantly faster readout system, an improved Plasma Interface, improved capabilities for unattended control, and a brand new software platform.

The new generation SPECTRO CIROS VISION ICP delivers a comprehensive range of new benefits to users including:

- new readout system—increases data processing speeds by a factor of ten, enabling real time standardization without loss of dynamic range, further improving precision and stability for the analysis of higher concentrations

- enhanced Plasma Interface—easier to handle and maintain while giving higher linearity
- optimized fluid paths—yielding shorter analysis times
- high sample throughput and 24 hour continuous operation
- extended control mechanisms, email alert function, and installation of WebCams for observation of the plasma offer excellent monitoring and diagnosis tools for unattended operation
- SMART ANALYZER VISION software platform – compliant to US FDA 21 CFR Part 11, the new 32-bit application offers unique reprocessing capabilities that enable elements to be added, method parameters to be edited and changed, even methods created and previously measured samples analyzed.

About SPECTRO

SPECTRO—the global leader in Optical Emission and XRF Spectroscopy—is a multinational company committed to innovation, instrumentation support and service. With more than 400 employees, SPECTRO has shipped over 20,000 instruments to customers throughout the world. SPECTRO is committed to producing advanced products, better and more diverse solutions, and providing an unparalleled support infrastructure. For more information visit www.spectro.com.

Nano & Micro Electromechanical Systems

Scientific innovator in high-resolution ultrasonic spectroscopy (HR-US) for material analysis will be unveiling its ground-breaking NEW high-resolution ultrasonic instrument, the Titration Kinetic Analyser on booth A4.220, Hall A4 at Analytica 2004, 11-14 May 2004 in Munchen, Germany. Ultrasonic Scientific is also showcasing its new customer-driven spectrometry software, HR-US accessories, and is holding on-booth seminars on how High-Resolution Ultrasonic Spectroscopy can be used for solving some of the main analytical problems facing the pharmaceutical, food, and personal care and polymer industries.

Ultrasonic Scientific's new Titration Kinetic Analyser will be launched in Europe at Analytica 2004. This new high-resolution ultrasonic instrument measures microstructural and molecular processes in liquids when concentration of selected compounds is changed (titration), or with time, when a chemical reaction is started by an injection of a required compound.

High-resolution ultrasonic spectroscopy is a novel technique for material analysis employing high-frequency acoustical (ultrasonic) waves. The application of ultrasound is well known; in medicine, ultrasound is used to visualize the internal organs of a patient's body. It is also used by submarines for underwater navigation.

Nanomaterials

Patented Process for Design, Development, and Production of Titanium Oxide Structures Customized for Titanium Metal Production

Altair Nanotechnologies, Inc. a developer and manufacturer of innovative nanomaterial products, was awarded its fourth patent entitled "Method for Producing Catalyst Structures" by the U.S. Patent Office. Patent number 6689716 was issued on February 10, 2004. The inventors of the process are Dr. Bruce J. Sabacky and Timothy M. Spitler, both of Altair Nanotechnologies.

"This patent, which further differentiates us within the rapidly growing nanotechnology market, underscores our commitment to bring nanotechnology out of the laboratory and into commercialization," said Altair President Dr. Rudi E. Moerck.

Titanium is the 10th most abundant element in the earth's crust and the fourth most abundant structural metal. Titanium metal has unique physical and chemical properties including high strength-to-weight ratio, good corrosion resistance and excellent ballistic armor protection while offering significant weight reduction over conventional steel and aluminum alloys. It is as strong as steel, but 45 percent lighter. It is 60 percent heavier than aluminum, but twice as strong. Titanium is expensive only because the current process for refining the ore to metal is a multi-step, high temperature batch process.

According to the AMPTIAC Quarterly, a Department of Defense-sponsored publication, current global production of titanium metal is approximately 50,000 tons per year at a market value of $600 million. AMPTIAC estimates that, due to the current state of manufacturing, titanium is produced at only about 1/20th of its current potential world volume. It is widely believed that a reduction of cost in the manufacturing process will expand the use to approximately one million tons.

The new patent describes the process for making microporous structures that can be used as a catalyst support or for electrode material. The microporous structures have a high porosity and high thermal stability, combined with good mechanical strength and relatively high surface area. In addition to the structures being used in a process to make titanium metal, the process is useful for making titanium dioxide for catalyst structures used for fuel cells, sensors, and electrochemical cells.

Nano & Micro Electromechanical Systems

Corporation has announced their intention to enter the spectroscopy and applied spectroscopy markets by launching the OSM Series of turnkey spectrometers, light sources and fiber based accessories. These feature-rich instruments incorporate multi-element detector arrays, optional fiber coupling and battery-operated portability, as well as a choice of detectors and gratings covering the UV through near IR. Higher end, self-contained models provide innovative options such as on-board 12-bit processing for data acquisition and real-time data analysis, on-board memory and removable SmartMedia memory cards for data storage and transport, simple touch screen control, and network connectivity via an optional Ethernet port. The OSM series is targeted at demanding laboratory research applications, process monitoring and QC labs, as well as OEM system-integrators.

Ron Hartmayer, Director of Marketing for Photonics Products, notes that, "Spectroscopic instrumentation is an obvious and natural progression for Newport on several fronts. First, we have long been supporting this market at the component level with both optics and high performance opto-mechanical assemblies. Second, all the core technologies deployed in these instruments are already strongly represented at Newport. Lastly, Newport has a history of supporting growing markets with value-added, integrated solutions that simplify the implementation of disparate photonic components. We certainly believe that spectroscopy and applied spectroscopy are markets with tremendous growth potential."

Hartmayer lists an extensive range of applications that the company intends to actively support. These include basic chemistry and biochemistry, biology and biomedicine, semiconductor research and fabrication, pharma, astronomy, agricultural products and food processing, and education.

More information on Newport's growing line of spectroscopy instrumentation is available on the company's Web site at www.newport.com.

Nanoelectronics

researchers Kathryn Guarini and Chuck Black made a splash on Dec. 9 at the IEEE's International Electron Devices Meeting in Washington, D.C., when they announced that they made a nanocrystal version of flash memory. But IBM is more bedazzled by the versatility of the fabrication technique than the device, according to Guarini. The flash application allowed them to prove the feasibility of their approach.

IBM uses a technique developed at the University of Massachusetts that allows polymers to self-assemble into a honeycomb pattern with holes as small as 20 nanometers. The polymers then are placed as a stencil on silicon dioxide, a material compatible with today's chip-making processes, for growing uniform nanocrystals.

Guarini said her research as a graduate student at Stanford University, where she specialized in maskless lithography techniques, taught her how to build nanostructures using probe tips.

The self-assembled honeycomb shape is one of several patterns Guarini said they want to create using various self-assembling materials. Different geometries would serve different functions as memory or logic devices. And their self-assembly system could complement other nanotechnology initiatives at IBM, such as efforts to construct transistors using carbon nanotubes.

Placing nanotubes into desired positions is one of many challenges for making a nanotube-based transistor. A simpler and faster approach is to grow the nanotubes where they need to be. That may be possible by placing a catalyst into a honeycomb chamber from which the nanotube could form.

IBM is in a race with companies such as Intel Corp., Hewlett-Packard Co. and Motorola Inc. to shrink components on chips for smaller, faster computing devices. The market for nanomaterials used in electronic devices will start in the billions and continue to grow, according to industry analysts at Business Communications Co.

Nanoelectronics

19, 2004 – When is a wire more than a wire? When you add the "nano" prefix, of course.

On one level, "nanowire" is a literal term: a channel for electrons or photons no wider than a few thousand atoms. As electronic components shrink closer to the size of biological cells or molecules, such wires may eventually need to replace the strands of metal that interconnect circuits today.

Nanowires are pinned under two electrodes.

Peidong Yang (News, Web) and colleagues at the University of California, Berkeley, have, for example, directed crystals of zinc oxide, a semiconductor, to self-assemble into arrays of nanowires that function as ultraminiature lasers and might be useful for probing cells or performing more-precise surgery.

Yang's group has also put a thin layer of silver nanowires to work as a delicate chemical detector. Such a film of nanowires could be a good way to sense chemical or biological warfare agents.

One of the most novel nanowires developed to date doesn't merely connect circuit components, but can work as an actual transistor or logic circuit. Charles Lieber's (News, Web) research group at Harvard fabricated such nanowires-as-devices by alternating bands of different semiconductor materials as the wire is formed.

The nanowires can then be applied at room temperature to the back surface of a display as an extremely thin coating. The wires lie flat and all point in the same direction, a little akin to a raft of logs on a river. Next, a grid of electrodes is deposited using conventional manufacturing methods on top of the nanowire film. The electrodes pin bundles of the nanowires across a gap to form the transistor elements that make up a display's pixels.

And because the process is low temperature, the nanowire film could be deposited on a flexible surface in a low-cost, high-throughput process such as "roll-to-roll," much the way newspaper is printed.

On Jan. 14, Nanosys announced it would work with Intel to investigate how its nanotechnologies might work in future memory devices. The company, however, declined to specify whether such efforts would focus on nanowires.

Nanomedicine

Calif., Jan. 5, 2003- Researchers have developed an improved method for performing sentinel lymph node biopsy, a crucial first step in determining whether a cancer has spread to other parts of the body. The new method depends on quantum dots, nanometer-sized crystals that emit near-infrared light, to illuminate lymph nodes during cancer surgery. The research, resulting from collaboration between researchers at MIT, Beth Israel Deaconess Medical Center and Brigham and Women's Hospital, will be published in the January issue of Nature Biotechnology.

The new near infra-red quantum dots were developed and synthesized at the MIT department of chemistry in the laboratory of Professor Moungi Bawendi, a scientific co-founder and advisory board member of Quantum Dot Corporation (QDC). The novel intraoperative, near-infrared fluorescence imaging system was developed in the laboratory of Dr. John Frangioni, Assistant Professor of Medicine and Radiology at Harvard Medical School and an Attending Physician at Beth Israel Deaconess Medical Center.

The study, entitled "Near-Infrared Fluorescent Type-II Quantum Dots for Sentinel Lymph Node Mapping," by Sungjee Kim, and colleagues, describes how the quantum dots were injected into live pigs and followed visually to the lymph system just beneath the skin of the animals. The new imaging technique allowed the surgeons to clearly see the target lymph nodes without cutting the animals' skin.

Sentinel Lymph Node (SLN) mapping, the surgical technique employed in the study, is a common procedure used to identify the presence of cancer in a single, "sentinel" lymph node, thus avoiding the removal of a patient's entire lymph system. SLN mapping relies on a combination of radioactivity and organic dyes but the technique is inexact during surgery, often leading to removal of much more of the lymph system than necessary, causing unwanted trauma. The current work was performed on laboratory animals, including near-human sized pigs, considered by scientists to be a good predictor of human results.

The study reported that the new imaging system with near-infrared quantum dots was a significant improvement over the dye/radioactivity method currently used to perform SLN for several reasons, including:

Throughout the procedure, the quantum dots were clearly visible using the imaging system, allowing the surgeon to see not only the lymph nodes, but also the underlying anatomy.

The imaging system and quantum dots allowed the pathologist to focus on specific parts of the SLN that would be most likely to contain malignant cells, if cancer were present.

The imaging system and quantum dots minimized inaccuracies and permitted real-time confirmation of the total removal of the target lymph nodes, drastically reducing the potential for repeated procedures.

"SLN mapping has already revolutionized cancer surgery. Near-infrared quantum dots have the potential to improve this important technique even further," said Dr. Frangioni.

"This is an impressive study," added Carol Lou, president of QDC. "While still several years away from being a reality for patients, this is a wonderful demonstration of the potential of quantum dot nanotechnology to significantly enhance medical care."

The paper can be found on the Nature Biotechnology website at: www.nature.com/cgi-taf/dynapage.taf?file=/nbt/journal/vaop/ncurrent/index.html

Nanomaterials

working to make microscopic and nanoscale machines and electronics have produced electrical wires at that scale, but it has proved more difficult to shrink the fiber optics that guide light. The trick to guiding light is finding ways to keep the photons confined to a fiber rather than leaking out.

Researchers from Harvard University, Zhejiang University in China and Tohoku University in Japan have made glass optical fibers as thin as 50 nanometers that guide light without losing much of it. Fifty nanometers is more than one thousand times finer than human hair. The researchers have made the thin optical fibers up to several centimeters long.

The key to such small, low-loss optical fibers was finding a way to make them very uniform in diameter and with very smooth walls, according to the researchers. To make the fibers the researchers first used a flame-drawing method to make micron-sized fiber. They then wound the fiber around a tapered, heated, 80-micron-diameter sapphire tip to keep the wire at a steady temperature while they pulled the fiber a second time to make it thinner.

The tiny optical fibers could be used in microphotonic devices for optical communications and optical sensing. The smaller fibers could lead to smaller and/or faster devices, according to the researchers.

Nanomaterials

are well known but the development of a new class of plastics that can be reshaped under pressure could lead to a new approach to manufacturing plastic products that avoids energy-guzzling high-temperature processing. US researchers devised the "ever green" plastics, which are also recyclable materials, to maintain their useful mechanical properties even after they have been remoulded.

Anne Mayes and her colleagues in the Department of Materials Science at MIT (Massachusetts Institute of Technology) in Cambridge have made the new class of plastics—baroplastics—from a nanophase blend of two polymers, polystyrene and poly(n-butyl acrylate) or polystyrene and poly(2-ethylhexyl acrylate). Under several hundred times atmospheric pressure, at 34.5 MPa, the clear solid softens and can be moulded into any shape desired. The idea of using raised pressure rather than temperature to process two otherwise immiscible polymer components is supported by small-angle neutron scattering (SANS) carried out by the team.

SANS reveals how low-temperature processing affects morphology too. Samples before and after room-temperature processing were analysed as well as after ten "recycling" cycles. The SANS measurements revealed how the rigid PS and soft PEHA mix together during processing, enabling the material to flow.

An additional advantage of using high-pressures as opposed to high-temperature for plastics processing is that the plastics themselves are not degraded by pressure in the way that they are by heating them even just to 200 Celsius. This means that the Mayes' nanophase polymer blends are almost limitless in the number of times they can be recycled without degradation. The team shredded and recycled their two baroplastics ten times and observed no obvious changes in molecular properties of the product as determined by preliminary mechanical and optical property testing and differential scanning calorimetry. Degradation is one of the serious limiting factors in current plastics recycling as it can destroy flame retardants and ultraviolet protectant additives as well as the polymers themselves.

"The baroplastics could provide alternative materials for applications currently employing semicrystalline polymers, rubber-modified plastics or thermoplastic elastomers, or as a matrix material for other thermally sensitive components such as biologically-derived therapeutics," Mayes told Spectral Lines. "The new materials should be compatible with much of the plastics-manufacturing equipment already in use," she adds, "which generally employ pressures of magnitudes comparable to those used in the present study."

Nanomaterials

Advanced Research Laboratory (ARL) is getting ready to commercialize a low-cost "nanostamp" technology for medical applications.

Hitachi's process creates "nanopillars" with extremely high aspect ratios (narrow relative to height), a feature that the company believes will prove useful for biochips and other applications, according to Akihiro Miyauchi, a senior researcher at Hitachi.

The technology uses a silicon "stamp" that presses onto a polystyrene-based polymer film, producing nanopillars that are extremely long and thin, about 3 microns in height. Right now, ARL is concentrating on making two versions of the nanopillars: one that is 250 nanometers in diameter and another that's 80 nanometers.

"With the 80-nanometer-diameter nanopillars, we are already producing pillars that are smaller than the current (90 nanometer) semiconductor process node by Intel or Fujitsu, for example, but we can easily go smaller because we have a very simple and very cheap process; it's just press and release," Miyauchi said.

He did not say how much smaller ARL plans to make the nanopillars, but he did say the process already works out to be much more cost-effective than semiconductor lithography processes.

Hitachi is not alone in developing very high-aspect ratio and market-ready process, according to Lars Samuelson, leader of the Nanometer Consortium at Lund University in Sweden. Lund's department recently developed techniques to direct self-assembly of nanowhiskers, he said.

This nano-imprint lithography (NIL) technique can make stamps with features of less than 20 nanometers in diameter, and "maybe close to 1: 100 in aspect ratio or even larger when optimized," Samuelson said.

Hitachi, meanwhile, is also gearing up for commercialization. Having developed the production technology, Hitachi has created a 15-member nanotechnology business enhancement office. Hiroshi Sonoda, senior engineer at Hitachi's nanotechnology planning office, said there is a $35 million market for the application in Japan alone after full-scale production. Hitachi said it plans to start shipping samples in the near future.

Neil Gordon, a partner at nanotech consultancy Sygertech and president of the Canadian NanoBusiness Alliance,said that development of low-cost products is key to applications like disposable medical tests. Hitachi, being a big multinational company, can afford to subsidize its nanopatterning products indefinitely, he said, giving the company "a considerable competitive advantage over dedicated nano-imprint lithography vendors."

To facilitate physical communication with the external environment, the vasculoid may be equipped with mechanical external ports which permit the ready attachment of cables or peripheral equipment as an alternative to purely wireless (e.g., radio, optical) communication links. Vasculoid ports include an appropriate macroscopic connector receptacle in hard contact with the sapphire structure. Ports may include connections to the internal plate-to-plate acoustic communications network or to docking bays and cellulocks to permit access to the internal material flow and thus allow inserting or extracting: (1) molecules and cells; (2) fresh vasculocytes and spare parts; (3) bioprovisions including respiratory gases, water and energy supplies; and (4) nanomechanical or other waste material. Extracorporeal user control interfaces could also be connected via an external port, if needed. Convenient and aesthetic locations for external ports include the navel or the nape of the neck. Respiratory gas and glucose transport requirements could be reduced by importing externally-supplied electrical energy—e.g., wall sockets (110 VAC at 0.9 amps equals the 100 watt human basal rate), backpack generators, solar collectors, nuclear batteries etc.—to be distributed throughout the appliance via wiring in the plates. These sources could be used to power: (1) local recycling of CO_2 and H_2O waste back into O_2 and high-energy carbohydrate fuel (mimicking photosynthesis energized by light), (2) the reconstitution of amino acids and proteins from urea and other nitrogenous wastes (as found in ruminants and in hibernating bears), and (3) other "reversible nutrition" processes, converting the body of the envasculoided user into a more closed-cycle system with reduced dependence on certain material and energy inputs. However, this approach would correspondingly increase user dependence on the chosen external power source and would require new in-plate or in-vesicle chemical processing plants without completely eliminating the need for any existing subsystem, hence may not be worth the added complexity.

Vasculoid installation leaves the user's gross thermal mass largely unchanged, as only ~4.4 kg of blood borne water (~9 per cent of the ~50 kg normally present in the human body) is removed and replaced with ~2 kg of sapphire vasculoid with a heat capacity equivalent to ~0.2 kg of water. Total passive aqueous heat capacity drops from 50.4 kcal/K down to 46.3 kcal/K. Nevertheless, 20th century patients with extensive metallic implants (e.g. pins, plates, bolts and joints) occasionally report brief chilly sensations during periods of cold weather and in other circumstances, due to the high thermal conductivity of metal compared with natural biomaterials or with plastics. Some organs such as the cornea are quite sensitive to sudden temperature changes—as little as a 0.3 K drop over a 0.785 mm^2 area for a duration of 0.9 sec (~700 nanojoules) is detectable by patients. A thermally conductive sapphire or diamondoid-coated sapphire vascular implant could extend these unpleasant sensations throughout the entire body, even for activities as simple as manually grasping a cold object. A targeted thermogenerative subsystem could help to eliminate this effect.

Perspiratory thermoregulation will continue as before. However, as explained earlier the active heat transfer mechanism of blood will be completely disabled. Normal capillary vasoconstriction/vasodilation mechanisms may be at least partially suppressed to avoid unnecessary tensions at the vasculoid-vessel interface. Detection of these responses, or direct temperature measurement, may be used to adjust passive thermal conductivity over a wide range, as follows. Each plate abuts neighboring plates through metamorphic bumpers with optimal thermal contact along stripe-like diamondoid buttons. Within each bumper, we may place an opposed pair of diamondoid pistons, separated by vacuum when they are pulled apart. When pulled apart, heat can be conducted only through the sapphire infrastructure, or vacuum in the plate, or through a surrounding aqueous medium, and is thus very slow, almost the same as normal tissue. But when the two pistons are pressed tightly together against diamondoid bumper buttons that are in good thermal contact, heat conduction largely bypasses the poorly-conducting sapphire regions and flows almost exclusively through the diamondoid contact region. Thus with pistons apart, the envasculoided human body has near-normal thermal conductivity; with pistons in contact, the body's thermal conductivity becomes near-metallic, perhaps as conductive as stainless steel. This transition is subject to user or program control, is switchable in microseconds, and may be directed only to specific volumes or pathways within the body if so desired*. A more complex system might employ thermal rectifiers, or, as J.S. Hall suggests, "heat pumps could be placed in the joints to make the whole phenomenon usefully controllable, augmented if necessary with tankers of ice and/or steam."

Installation of a full vasculoid appliance permanently displaces ~4.2 liters of natural blood volume, freeing up ~1 gallon of internal storage volume. At least some of this volume may be used for containerized temporary caching of consumables including surplus glucose, fats, water, oxygen (e.g. a high-pressure nanolung), minerals or other useful biomaterials, or various bio- and nano-wastes. For example, a 1-liter 1000-atm O_2 solid-walled cache holds ~34 hours of oxygen at the human basal metabolic rate or ~2 hours at the maximum rate. The free ~4.2-liter volume could also be used to store a wide variety of useful equipment or tools including computers, computer memories, external communications or navigational devices, solar energy accumulators, weapons, spare vasculocytes and other special purpose nanorobots, spare parts, or useful tools. An entire spare vasculoid (~558 cm^3) could even be stored in this volume.

The vasculoid, like the micron-sized respirocytes proposed elsewhere, will offer the ability to breathe at low O_2 partial pressures. The vasculoid gas tanker fleet can hold ~20 minutes of oxygen at the basal metabolic usage rate, or up to ~100 minutes if most of the tanker fleet is diverted to respiratory gas carriage in lieu of other applications. Vasculoid installed in underwater divers who are breathing pressurized air to modest depths should allow only enough nitrogen into the body to forestall mechanical tissue damage, then rotor it back out again as the diver surfaces to avoid decompression sickness. At 100 per cent saturation the body absorbs ~10^{21} excess N_2 molecules per meter of depth. Each tanker could hold ~7.98×10^9 nitrogen molecules (at 1000 atm), requiring the storage capacity of 0.125 trillion tankers per each 1 meter of decompression. Given a reconfigurable fleet of ~166 trillion tankers, expedited decompressions from ~100 meters are probably achievable using the present design. To establish a greater vasculoid operating depth, in theory an auxiliary 1000-atm storage tank "nanolung" installed in the vasculoid wall could provide ~9.43 cm^3 N_2 storage capacity per each 100 meters of decompressible diving depth. However, nitrogen should not be used to hyperpressurize the tissues due to nitrogen narcosis. It should also be noted that obesity is a risk factor in decompression sickness because nitrogen is 5 times more soluble in fat than in water and because of reduced blood access to adipose tissue, and also compression and decompression are asymmetrical: a nitrogen load acquired in a few minutes may take hours to fully deplete purely by diffusion (and the extraction rate varies markedly by tissue type), so extravasculoid respirocyte-class nanorobots may be required for optimal results. Clearly the vasculoid can achieve a decompression rate comparable to breathing pure oxygen, but without the oxygen toxicity. Helium may be useful for mixed-gas diving, but is prone to leakage some will be lost from the body during use and cannot effectively be replaced from the environment.

The lymphatic system extends into all the same tissues as capillaries and is structurally similar to the venule network. However, since lymph vessels constitute a "cul-de-sac" system, not a "circulatory" system, it is unnecessary to install vasculoid systems in the lymphatics in order to achieve most of the positive benefits expected from an arteriovenous vasculoid. With the circulatory vasculoid in place, the workload of the natural lymphatic system may be greatly reduced—protein leakage from capillaries, entry of particulates into the tissues, and the presence of pathogens requiring immune system response all should be greatly reduced. More properly, lymphovasculoid should be regarded as optional equipment which may permit more precise control of the immune system in general and of lymphocyte traffic in particular. Total adult lymph flow is typically ~2 liters/day, so transport requirements within the lymphovasculoid should not be particularly severe. If implemented, the lymphovasculoid would be emplaced as two distinct installations—a right lymphatic and a thoracic subsystem—following the natural division in the human body. Special lymph-recycling facilities may be required at the locations where the thoracic duct and the right lymphatic duct join the venous tree.

Pathogen disposal sites using an enzyme-based digest and discharge protocol may be located near lymphoid tissues and organs (to permit immune system processing) and near excretory organs such as kidney, liver, gallbladder, and gut (to facilitate removal from the body). Specialization of this function at specific locations appears more efficient than on-site pathogen processing systems which would need to be numerous, widely distributed, and usually idle.

It may be useful to deploy sensor probes which can leave the vasculoid and enter the surrounding tissues to detect and monitor remote events (such as angiogenesis, tumors, or pathogens) that otherwise might not be conveniently detected from within the vasculoid. Additionally, large blocks of tissue are only lightly vascularized or have no capillaries whatsoever, as for instance the synovial chambers in skeletal joints and the epidermis. Medical nanorobots capable of migrating through noncellular tissue might be useful both for repair purposes and for maintaining a disease-free condition in these tissues. Extravasculoid devices may also constitute general-purpose mobile cell repair machines with even broader injury response and prevention capabilities.

Although the vasculoid itself is essentially fireproof (sapphire and ruby will not burn in oxygen), the overlying human tissue is not. Thus another optional upgrade to the basic vasculoid package is an active damage control subsystem that offers some protection from severe thermal burns. This optional subsystem would include installation of biocompatible sensor-tipped diamondoid or sapphire pores that pass from the epidermis through the dermis to the nearest envasculoided capillaries. The comparative human skin response to sapphire vs. fluid cooling has been studied experimentally. Upon detection of a potentially harmful thermal event, sacrificial water is pumped from internal reservoirs into channels in the plates, and from there into the pores, eventually emerging as billions of aqueous microstreams. Touching a red hot object stimulates a nearly instantaneous emission of ablative water, producing a protective layer of heated water and steam between the hot object and the skin. Protection of the human hand (~150 cm^2) up to ~1300 K (~decomposition temperature of diamond) for a ~1 second exposure requires a 162 kilowatt/m^2 heat dissipation rate and a ~1 cm^3 sacrificial water reservoir. Of course, this system could easily be overwhelmed if the user is exposed to intense IR radiation, such as inside burning buildings or near large explosions, and in any case the hair is not protected as well as the skin.

The overall static structure of the basic vasculoid appliance is an array of rigid plates, strongly fastened together, entirely covering a curved two-dimensional surface embedded in a three-dimensional space. The plates conform to ordinary movement, but may adequately resist certain kinds of destructive movement such as separation due to cutting, rapid accelerations or decelerations, and crushing injuries. Although local endothelial cells may be damaged during such sharp movements, an envasculoided tissue should be somewhat more resistant to gross damage. For example, the critical buckling pressure of a hollow tube of circular cross-section deformed into an elliptical cylinder by external pressure is given by Freitas as $p_{crit} = (E\, h_{tube}^{\;3})/(4\, r_{tube}^{\;3}\, (1 - c_{Poisson}^{\;2}))$ where E is Young's modulus (~10^6 N/m^2 for vascular tissue and ~10^{11} N/m^2 for sapphire, h_{tube} is tube wall thickness (~1 mm for aorta, ~ 1 mm for capillary and for vasculoid plate-tube, r_{tube} is tube inner radius (~25.0 mm for aorta, ~8 mm for capillary, ~7 mm for vasculoid plate-tube; and $c_{Poisson}$ is the Poisson ratio for the material (~0.3 for vascular tissue, ~0.1 for diamondoid). Using this formula and the stated values, p_{crit} for aorta is ~59 N/m^2 and the vasculoid coating adds only negligible additional resistance to crushing, p_{crit} ~ 0.002 N/m^2. However, for the ubiquitous capillaries p_{crit} ~ 540 N/m^2 but adding the vasculoid coating can dramatically increase crushing resistance up to ~7 x 10^7 N/m^2, or ~700 atm of overpressure. Since the mechanical coupling between plate bumpers is subject to both design and conscious user control the crushing resistance of capillary beds can be autogenously varied over five orders of magnitude. Of course, high overpressures may seriously damage the underlying natural endothelium but this might nevertheless be acceptable over small areas during emergency situations.

For instance, resistance to deep incision-slash wounds would be markedly improved. Crushing strength may be increased using thicker plates or stronger bumpers.

Similarly, the bending stiffness of a hollow tube is given by Freitas as k_{shaft} = 3p E (R_{tube}^4—r_{tube}^4) / (4 L_{tube}^3), where the outside tube radius is R_{tube} = r_{tube} + h_{tube} and the tube length is L_{tube} (~400 mm for aorta, ~1 mm for capillary. Using this formula and the stated values, the resistance to bending (stiffness) of the aorta increases 60-fold (to ~230 N/m) when coated with a single layer of vasculoid plates, but the stiffness of capillaries increases nearly 70,000-fold (to ~0.4 N/m) when coated with diamondoid plates. The buckling strength of coated vessels compared to natural vessels may exhibit similar orders-of-magnitude improvement, though in practical systems some of this potential will be lost because interbumper connection rupture strength may be only 10^7-10^9 N/m^2, 1-3 orders of magnitude less than for solid diamondoid materials. The natural musculature may be too weak to move soft tissues that are thoroughly penetrated with ultrastiff capillaries, but selective autogenous control of plate bumper expansion and contraction should make possible any necessary macroscale voluntary movements, or on-demand mechanical rigidification of specific limbs or organs. Bumpers may be driven at ~KHz frequencies, permitting positional adjustments on millisecond timescales. Such motions will require additional energy expenditure.

A complete examination of the acceleration tolerance of the envasculoided whole human body would have to take into account a wide variety of factors including differential density of body parts (e.g., lung, soft tissue, skeleton), magnitude and direction of the stress vector (e.g., positive or negative, axial or transverse, etc.), duration and timing of the stress vector (e.g., acute or chronic, linear or periodic), rate of onset and pulse shape of the acceleration, and the mechanical characteristics both of linked vasculoid components and of the vasculoid-endothelial interface. However, a simple order-of-magnitude estimate may be cautiously ventured, as follows. A vasculoid appliance of differential density Dr_{vasc} relative to soft tissue and mean thickness h_{vasc} in contact with biological tissues that are subjected to a linear acceleration of a = (a_G g), where g = 9.81 m/sec^2, pushes the vasculoid into the tissues with a pressure force of P_{vasc} ~ h_{vasc} Dr_{vasc} g a_G. Because the vasculoid tube has 2 opposed walls, each 1 mm thick, and because the tanker fleet covers 4 per cent-55 per cent of the surface of each of these opposed walls with tankers that are 1 mm thick, then the effective capillary thickness of pushed vasculoid under acceleration ranges from h_{vasc} ~ 2.1-3.1 microns. Tissue density is r_{tiss} ~ 1050 kg/m^3; sapphire tanker density is r_{tank} ~ 993-1738 kg/m^3 if filled with vacuum or water at 310 K, and plate density is r_{plate} ~ 2000 kg/m^3, giving an effective vasculoid density of r_{vasc} ~ 1643-1988 kg/m^3 under various usage conditions. Hence Dr_{vasc} = r_{vasc}—r_{tiss} ~ 593-938 kg/m^3 and so the maximum tolerable G-force for envasculoided tissue is a_G ~ P_{vasc} / (h_{vasc} Dr_{vasc} g) = 24.4 P_{vasc} ~ 4880 G, taking P_{vasc} ~ 200 N/m^2 as the tentative limit for acute vascular damage because this has been found experimentally not to cause gross injury or denudation of canine arterial endothelium by shear force.

However, endothelial transcription factor responses (protein expression) can be triggered by shear forces as low as 0.02-2 N/m^2 (a_G ~ 0.5-48.8 G) and normal physiological blood shear forces are 0.14-2.6 N/m^2 (a_G ~ 3.4-63.4 G), so the maximum chronic external acceleration indefinitely tolerable by an envasculoided human body probably may not exceed 50-100 G without producing altered, possibly pathological, cytochemical states. (The operational limit of the current vasculoid design is also ~30-100 G). This still would represent a substantial improvement over the natural acceleration tolerance of the unaided human body—the mean relaxed tolerance is ~3.23 G and loss of brain blood flow occurs at $+G_z > 4.5$ G. Consciousness has been retained for a maximum of 45 seconds at 9 G using G-protective equipment and straining maneuvers developed for the U.S. Air Force, and for 4 minutes at 12 G or 4 seconds at ~16 G using water immersion. Severe impact injury to humans occurs from $+G_z$ accelerations of 30 G exceeding 100 millisec or 100 G exceeding 2 millisec—the recordholding primate is evidently one of Colonel Stapp's chimpanzee test subjects that survived ~1 millisec of 247 G in the $-G_x$ direction on a rocket sled, suffering only "moderate" injuries. Thus, the basic vasculoid may improve maximum human acceleration tolerance by a factor of 10-20, or more, and higher-G-tolerant architectures can probably be devised. With sufficient capability, the vasculoid structure could allow the non-local distribution of stress—for example, by spreading the force of an impact in a manner similar to a bulletproof vest, also helping to quickly close any vascular breaches that might occur. A common and significant personal injury mechanism is brain-skull impact, and a tough interwoven vascular support scaffolding could help to fix the brain within the skull, potentially reducing the danger of concussion for at least moderate decelerative loads. (Active suppression of rotational impact trajectories may enhance cerebral durability, as demonstrated using high-speed cinematograph films of woodpeckers which regularly survive 6-7 m/sec cranial impact velocities with ~1000 G decelerations.) Positional information and selective sectioning could allow unavoidable partitions of tissue to occur along straight planes, producing a minimum of damage when compared with ordinary tearing injuries. Whether the vasculoid appliance would be a liability in explosive overpressure situations, where a fluid-filled vasculature might not (e.g., lethal effects in humans noted for 40 psi to 50 psi overpressure shockwave; 300 psi (~20 atm) record for successful deep water submarine escape), deserves further study. A preliminary scaling study for a conceptual design of a single, complex, multisegmented nanotechnological robot that appears capable—with numerous caveats, as noted—of duplicating all essential thermal and biochemical transport functions of the blood, including circulation of respiratory gases, glucose, specialty biochemicals, waste products, and all blood borne cellular elements. The vasculoid, a 2-kg ~200-watt intimate personal appliance, conforms to the shape of the existing vasculature and may serve as a complete replacement for natural blood while greatly improving the durability and functionality of the human body. The device described in these chapters would represent a most extreme intervention using a very advanced medical molecular nanotechnology. It also should be noted that our current knowledge of the biological functions of the circulatory system is incomplete, so the design presented here must be considered provisional at best. But the principal challenge was to advance a plausible argument that a nanomechanical whole-body thermal and biomaterials transport system would violate no known physical, engineering, or medical principles, could presumptively be made adequately safe for the user, and might confer some significant advantages over simpler whole-body systems exclusively employing unlinked populations of individual blood borne and tissue-borne nanorobots.

Ultimately, and from the standpoint of human-guided evolution, the body exists primarily to ensure the survival of the mind—not the replication of the genes, which was the ancient paradigm. It would seem that a somewhat more advanced and compact version of the proposed device could function independently of nearly all noncortical tissue. Thus the vasculoid is most fascinating because it may represent one last outpost of humanity at the final frontier of biological evolution.

Applications and Importance Nano-Tribology

The term tribology is derived from the Greek word *tribo* meaning rubbing and *logy* meaning knowledge. The original applications by the Greeks of tribology were in trying to understand the motion of large stones across the earth's surface. Today tribology has grown to include the methodical study of friction, lubrication, and wear. Tribology plays a critical role in diverse technological areas. In the advanced technological industries of semiconductor and data storage, tribological studies help optimize polishing processes and lubrication of data storage substrates. In traditional industries such as automotive and aerospace, tribological studies help increase the lifespan of mechanical components. Many industrial processes require a detailed understanding of tribology at the nanometer scale. The development of lubricants in the automobile industry depends on the adhesion of nanometer layers (mono layers) to a material surface. Assembly of components can depend critically on the adhesion of materials at the nanometer length scale.

There are a number of traditional tools for characterizing friction, lubrication and wear. The most common characterization tool is the tribometer having several configurations such as pin-on-disk, ball on flat, and flat on flat, etc. Generating motions at the nanometer scale is extremely challenging. New characterization techniques are required to understand tribology at the nanometer scale. The atomic force microscope is now being routinely applied for studying nanoscale tribology. The natural extension of the AFM for tribology applications is derived from the motion of a nanometer-sized stylus in the AFM over a surface. Although traditional tribology testing is not done with an AFM, many new types of applications are possible.

Examples of the application of AFM to tribology include:

- Direct three-dimensional visualization of wear tracks, or scars on a surface.
- Measurement of the thickness of solid and liquid lubricants having nanometer or even monolayer thickness.

- Measurement of frictional forces at the nanometer scale.
- Surface characterization of morphology, texture, and roughness.
- Evaluation of mechanical properties such as hardness and elasticity, and plastic deformation at the nanometer scale.

A major advantage of the AFM for tribological studies is that the AFM can be routinely used on all types of materials. Materials commonly studied include: ceramics, metals, polymers, semiconductors, magnetic, optical, and biomaterials. AFM investigations are usually made in ambient air environment. It is possible to make AFM studies in a vacuum or liquid environment.

The effects of wear at the nanometer scale become critical to the optimization and stability of machines as the tolerances in precision machines become smaller and smaller. Traditional microscopes such as the optical and scanning electron microscopes facilitate visualization of wear in 2-dimensions. For example, with the SEM it is possible to get a magnified view of wear tracks in the x-y axis but cross sectioning is required for measuring the depth of wear tracks. The AFM allows direct 3-dimensional visualization of wear tracks and scars. The images may be displayed in a 2-D projection and a 3-D projection. Direct measure of wear track depth can be easily measured with a line profile derived from the AFM image.

It is well known that layers of lubricants on surfaces that are less than 100 nm can dramatically affect lubrication behavior. Characterization of such films is necessary for developing optimized lubricating films. However, nanometer scale characterization of lubrication films offers a substantial challenge. Optical techniques such ellipsometers can be used for measuring lubrication thickness of large sections, (greater than 10 square micrometers), of a surface. Measurement of the localized (less than 1 micron) film thickness is not possible with the ellipsometer. The probe is mounted at the end of a cantilever in an AFM making it possible to measure interaction forces between the probe and the surface by monitoring the deflection of the cantilever. A graph, called a force/distance curve, shows the forces on the probe as the distance between the probe and the surface are reduced. The nature of the force/distance curve depends on the force constant of the cantilever, the lubrication density, probe geometry, and the lubrication thickness. By measuring the changes in force/distance curves in an AFM it is possible to directly ascertain the thickness of lubrication films. Below is an example of a force/distance curve for a surface with no lubrication film compared to one with a lubrication film. The thickness of the film is established from the force/distance curve.

Friction between two surfaces depends on the chemical and mechanical interaction between the surfaces. Changes in chemical composition giving rise to friction are measurable with the AFM. The technique for measuring these forces is called lateral force, or frictional force microscopy. As the probe moves over a surface in the AFM, changes in the chemical composition of the surface can give rise to torsions of the cantilever on which the probe is mounted. The torsion of the cantilever is then proportional to the friction between the probe and the surface. In an AFM it is possible to simultaneously measure topography and frictional force images. The topography image is derived from monitoring the vertical forces on the cantilever and the friction image is acquired simultaneously by monitoring the lateral motions of the cantilever. Below is a FFM image of a sample illustrating changes in the friction.

The AFM gives extremely high contrast on surfaces that are flat at the nanometer scale. Optical and electron microscopes are not able to resolve surface texture that is easily measured with the AFM. Applications include the visualization of surface topography in 2-D and 3-D perspectives, line roughness measurements, and area roughness measurements. All of the traditional area and surface roughness parameters can be calculated after the AFM image is acquired.

Mechanical properties such as hardness, elastic modulus, stiffness and compressibility as well as material behavior such as plastic deformation, and fracture can be studied with the AFM. It is possible to study nanohardness by directly pressing an AFM probe into a sample's surface; however, it is advantageous to use an instrument that is optimized for nanoindentation. The primary advantage of the nanoindenter over an AFM for nanohardness measurements is that it is easier to get calibrated measurements with the nanoindenter. It is useful to use the AFM to measure the three-dimensional topography of indentations made with a nanoindenter. AFM images allow direct visualization of material deformation or fracture behavior.

Using techniques such as pulsed force mode, the stiffness of a sample at a matrix of locations is measurable. From this data it is possible to create a stiffness mapping of a surface. Stiffness maps can only be made on samples where the stiffness of the surface is lower than the stiffness of the cantilever. Such stiffness images are routinely measured on polymer samples. Adding a fixture to the stage of the AFM makes the study of material behavior such as plastic deformation and fracture possible. The fixture permits creating forces on a sample while AFM images are being taken. A variety of materials may be studied with such a technique.

Tribiology (friction, lubrication, and wear) is the science of interacting surfaces in relative motion. The word "tribology" is from the Greek *tribein*, meaning to rub. Although the term may be unfamiliar to many, its meaning is clearly conveyed among some of the oldest written records. An Egyptian painting dating back to 1880 BC depicts workers dragging a sled containing a heavy statue. One worker pours a liquid on the ground just before the runners to make the going easier. Leonardo da Vinci addressed the same problem when he wrote in 1519, "All things and everything whatsoever however thin it be, which interposed in the middle between objects that rub together, lighten the difficulty of this friction." An excellent source for further historical reading is *Dowson's History of Tribology*. We have first-hand experience with friction any time we walk on ordinary ground and compare the outcome with what happens when we try to walk on ice. Wear of mechanical parts—as every car owner knows—eventually leads to failure and is one of the most costly problems facing industry.

To design and engineer mechanical parts, we need a macroscopic understanding of tribological processes. However, today the tolerances of many components, such as optical devices, computer chips, and ultra-smooth surfaces, are approaching the microscopic (atomic) length scale. In fabricating such precise components, we can greatly benefit from an atomic scale understanding of what happens when surfaces interact. The emerging field that lets us obtain this atomic scale understanding of fundamental processes of surfaces in motion is called molecular tribology or nanotribology. (A nanometer is one billionth or 10Ð9 of a meter; the size of an atom is about 0.3 nm.) Recent progress in nanotribology is being driven by advances on several fronts. Of primary importance is the advent of probes that allow researchers to examine interacting surfaces at a microscopic level. One such probe is the atomic-force microscope, which has a very small tip with a radius of about 10 nm used to probe a surface. The surface force apparatus has also become a highly useful tool. Other major developments in the last decade are better models of interatomic forces, which, coupled with improved performance of computers, are giving researchers a new window into tip—surface interactions through computer simulations.

The molecular dynamics method was originally developed in the late 1950s to study the statistical mechanical properties of a collection of atoms at equilibrium. The method is now a well-established and important tool in the fields of physics, chemistry, biology, and engineering. Recent advances in computer hardware and software allow us to simulate relatively large systems using increasingly realistic models of the forces between atoms.

In principle, MD modeling is simple. We start by identifying the position and velocities of all atoms in a system we want to model. Then we calculate the force on every atom from its neighbors and advance the positions of the modeled atoms according to Newton's equations of motion (force = mass × acceleration). During a simulation, we evaluate the response of a material—often copper, silicon, or silica glass—as it is subjected to an external force by following the response of each atom. Examples of external forces include stress fields, moving boundaries, and heat baths. Despite the simplicity of MD modeling, the number of users and applications today remains limited; in addition, many researchers do not use it to its full potential. The reason for this under use is that this type of modeling can be "overkill" in that it is too expensive and requires too much detail for solving most practical problems. Furthermore, there is a certain art in selecting the appropriate boundary conditions and constructing interatomic-force models.

The boundary conditions used in many of the simulations of indentation and cutting of a clean metal surface by a diamond-like tool. In this figure, the dashed line forms a box surrounding all the atoms that are represented. This box is called a simulation cell, which is the "window" on the system being modeled. Both the metal surface and the V-shaped diamond tool above it have "handles" to position the surfaces relative to one another. These are the dark circles forming the boundary. In practice, the tool is fixed in the rotational direction and slide the work surface from left to right during a simulation of cutting. Immediately next to the boundary atoms is a region in which the atoms are constrained to be at room temperature. From an atomic point of view, temperature is a manifestation of the vibrational energy of the atoms. The role of the thermostat atoms in the tribology simulations is to draw away heat generated at the tip of the tool, mimicking a much larger piece of material than is represented in the illustration. The remaining atoms represented by open circles make up the Newtonian region, meaning that these atoms are free from further constraint and move according to Newton's equations. The tribological processes in this region are of interest in the MD simulation.

In many cases, we would like to simulate cutting or scraping over lengths that are far greater than the dimensions of the dozen or so atoms. However, as mentioned earlier, following the motion of a large number of atoms is prohibitive in terms of computer time and cost. To help overcome this problem, a relatively simple approach is used. Atoms are allowed to leave the simulation cell at the "downstream" end (to the right as the simulated work piece moves under the tool). At the same time, periodically new atoms are inserted from the left at the upstream boundary. This approach restricts the population of atoms simulated at any given time, while allowing to model nanotribological processes in a much larger work piece. The extension of this model from two to three dimensions is straightforward.

One of the first simulations we refer to here involved the indentation of a metal surface by a blunted, triangular diamond tip moving vertically at a constant velocity ranging from 1 to 1000 m/s. Keep in mind that in this and all the subsequent work talked about here, the objective is to understand—at a more microscopic, atomistic level than was previously possible—what is happening when surfaces in relative motion interact with one another. A "snapshot" of what happens when the tool indents the first three layers of metal. The interatomic forces in a metal arise from two effects: the interaction between electrons localized on the atoms, and the interaction of atoms with free electrons. An embedded-atom model of the metal is used that incorporates both of these effects. In this figure, the atoms are shaded by the local value of stress. At first, the surface responds elastically (deforms without permanent loss of size or shape), and we see circular regions of constant stress, called the Hertzian stress field, which is well known from elasticity studies of contact mechanics. When the tool is pushed in further to indent six layers of metal, the surface begins to flow plastically, and we obtain shear and discontinuity. In a 2D simulation like this one, dislocation edges are clearly visible within the metal along the darker colored slip bands that appear at 60° angles with the surface. In a close-packed metal such as this, the atoms behave like hard spheres (for example, billiard balls), and slip is analogous to sliding stacks of billiard balls over each other A 3D simulation of the indentation of copper gives a slightly different picture. An image obtained after copper was indented at a rate of 1 m/s, and the simulated tool was withdrawn from the surface. (Slower rates, typical of experiments performed in a laboratory, are beyond the current capabilities of MD simulations.) In a 3D simulation, we do not see the distinctive bands of dislocation that are observed in a 2D simulation. Moreover, there is relatively little distortion at the metal surface. Instead. we find a small pileup of atoms on the metal surface, a few atoms in interstitial positions (between regular lattice positions), and a small dislocation loop.

To explain the difference between the 2D and 3D simulations. Here, the atoms are colored according to the initial layering before the computer experiment. This cross section at the moment of full indentation of the copper surface shows that very little surface distortion has actually occurred. Remarkably few atoms have bulged out around the tool. Instead of distinctive dislocation bands, we see disorder only within a few layers of the sides and bottom of the tool tip. This localized disorder arises from point mechanisms (interstitials, vacancies, and so forth) of plasticity.

There are many other ways to display the results for such a simulation. One of the best and most revealing is by means of a graph, called a loading curve. The load (instantaneous force in nanoNewtons) on the diamond-like tool as a function of depth of indentation (shown here as the number of layers indented) for the simulation. The very rapid fluctuations in load—which look like "noise" on the curve—arise from the rapid motion of surface atoms repeatedly colliding against the tool atoms. The larger peaks and valleys are far more interesting in terms of what they reveal. At first, the load rises linearly as the metal surface responds elastically. After indenting about 1.5 layers of copper, the load drops abruptly when the elastic stress is relieved. This abrupt drop, called "critical yielding," is reminiscent of what is observed in the laboratory. The first yielding in the simulation corresponds to a single copper atom popping out onto the surface from under the tool tip and relieving the stress energy. With more indentation, more of these point events take place. After indenting seven layers of copper, the tool is stopped. At this point, the surface accommodates nearly all the tip through elastic and plastic deformation, and there is little pileup of atoms around the tip, as observed in the cross section. When the direction of the tool is reversed (lifted from the surface), the load quickly drops to zero, as expected. However, the load suddenly rises once again as the tool is removed further. This unexpected rise is caused by annealing at the surface of the metal so that the material returns to more intimate contact with the tool, and the load rises.

The results from simulations using silver as the indented material are quite similar to those just described for copper. However, because silver is softer, the yielding occurs at a smaller load than for copper.

Cutting is a widely used process in fabricating components. To obtain a more basic understanding of what happens when a metal is cut, a simplified geometry is used. We idealize the model slightly by limiting the motion to two dimensions or thin, periodic slabs in three dimensions. This idealization, known as orthogonal cutting, assumes that the material being cut is sufficiently wide that edge effects have essentially no influence on the results. Orthognal cutting is stimulated in both two and three dimensions. The 2D simulations more accurately model the length scales and time scales that are common in laboratory experiments using single point, diamond-turning machines. In this type of simulation, cutting speeds range from about 10 to 100 m/s. We have also varied the sharpness of the rigid tool tip used for cutting. We have used tool-edge radii of curvature ranging from 1 to 20 nm. The bluntest tip (20 nm) approaches the sharpest real-life tool tip, which has a radius of curvature of about 35 nm. A single frame from a computer-animated movie of the three-dimensional MD simulation of orthogonal cutting. The copper material flows from left to right at 100 m/s. In this simulation, the carbon atoms in the diamond-like tool are not permitted to move, so they serve as the fixed frame of reference.

This type of projection makes it easier to see some important effects. During our simulation of cutting, the system forms a chip. The chip is reminiscent of what is actually observed during real experiments. The chip remains crystalline, but it has an orientation different from that of the surface. Regions of disorder among the atoms are apparent in front of the tool tip and on the surface in front of the chip. Unlike the 3D studies of point indentation, dislocations readily form during orthogonal cutting. Unlike the single point in the indentation simulations, the cutting tool in the orthogonal cutting geometry forms an infinite line, like a knife edge. This geometry provides much more energy for the creation of dislocations.

The calculations show that the cutting force strongly depends on the sharpness of the tool. This relation is also observed experimentally. Dull tools (those with a large tip radius) require larger forces to achieve the same depth of cut as sharp tools. A good measure to quantify the observations is the work (force x distance) performed by the cutting tool divided by the volume of material removed, also known as the specific energy. This energy increases dramatically with decreasing depths of cut. For our shallowest (nanometer scale) cut, the specific energy exceeds the energy required to vaporize the material, though the material remains a solid. The dependence of specific work on depth of cut has been observed in macroscopic metal cutting for many years and is known as the size effect. However, the macroscopic size effect is much less dramatic than the size effect observed in the simulations or in experiments using single-point, diamond-turning machines. The transition between macroscopic and microscopic behavior occurs at a length scale of a few micrometers, comparable to the average grain size in most metals. This result is interpreted as a change in the mechanisms of deformation, from grain-boundary sliding and motion of existing dislocations at the macroscopic scale to the creation of new dislocations and other point mechanisms of deformation at the microscopic, atomic scale. These dislocation-creation and point mechanisms of deformation consume significantly greater energies, leading to the observed size effect. In contrast to copper, materials like silicon and iron are not considered machinable with diamond tools. The reason is that the diamond tool wears rapidly, and it becomes difficult to maintain contour accuracy. The increased wear occurs for all materials that form strong chemical bonds (covalent bonds) with carbon. We have investigated and identified some of the underlying processes that give rise to these effects.

In covalent materials like silicon, the nature of interatomic forces is much different from that in metals. The strength of a covalent bond depends on the local environment, and the material forms an open structure. The basic method is the same as that for cutting copper, except that now we evolve not only the silicon surface material but also the carbon atoms in the diamond tool to allow for tool wear. It has been found that both copper and silicon show ductile behavior during cutting; however, the underlying mechanisms that allow this behavior are very different for the two materials. Whereas a copper chip remains crystalline, a silicon chip is transformed into a completely different state.

This time of a silicon-cutting simulation. In this case, the diamond wedge is cutting at a speed of 540 m/s. Although no wear of the tool is apparent yet, a single layer of silicon atoms has coated the diamond tip. Chip formation occurs between this layer of silicon atoms and the crystalline surface. In related simulations of diamond asperities (abrasive tips) scraping a silicon surface (as in grinding processes), it was found that the diamonds wear by forming small clusters of silicon carbide, which remain on the silicon surface. The silicon material in the chip and in the first few layers of newly cut surface appears to be amorphous or possibly to have melted. The temperature in the chip, calculated from the vibrational motion of silicon atoms, is comparable to the melting temperature of bulk silicon. However, the silicon atoms in the chip are not diffusing, which demonstrates that the material is, indeed, in a solid state. The following idea has been developed to explain chip formation in crystalline silicon. Because of the nature of covalent bonds in crystalline silicon, the energy needed to shear the crystal is enormous. In fact, less energy is required to transform the crystal into an amorphous solid and to shear the amorphous solid than to shear the crystal. The interplay between the energetics of different deformation processes and the dependence on length scale may explain the transition from brittle to ductile behavior that is observed in ceramic materials.

The ability to precisely machine ceramic surfaces like silica glass affects many different fabrication technologies and is central to constructing a wide range of sophisticated optics systems. A laser program at a University needs to make very high-quality mirrors, and we need to know more about special materials, such as ceramics, used in modern devices and equipment. We are applying MD methods to examine deformation processes at the atomic scale in these materials. The main difficulty is that glass and ceramic materials are brittle. Anyone who has seen a rock impact a windshield or a pane of window glass knows about the property of brittleness. This property can result in cracking, subsurface damage, and other kinds of costly damage during precision machining. However, a growing body of experimental evidence suggests that glass and ceramics can exhibit both ductile and brittle behavior during grinding, depending on the size of the abrasive used. In particular, there is an abrupt change in the surface smoothness of glass with a smaller abrasive size. In other words, with a smaller depth of cut, even glass can behave in a ductile manner. If we could better understand and manipulate this ductile-to-brittle transition, we could improve the economics associated with fabricating ceramic components. To address this problem, simulation of fused silica glass with no impurities was undertaken. In silica, four oxygen atoms surround each silicon atom to form a tetrahedron. Each oxygen atom is shared with two silicon atoms to form connected tetrahedra. Amorphous silica glass is a random network of these interconnected tetrahedra. To study the nanometerscale deformation of fused silica, a diamond tip was pushed with a radius of about one nanometer into a smooth silica surface.

It was found that the silica surface responds much more elastically than did either the copper or silicon surfaces. For indentations up to 1.25 nm, it was observed that no plastic deformation occured. At an indentation of 1.25 nm, a significant rearrangement of the silica network occurs directly beneath the tool tip. This rearrangement, an example of which is shown, leads to a decrease in the slope of the loading curve. When the tool is removed, the unloading curve does not follow the loading curve, as it does for elastic indentations up to 1.25 nm. The area between the two curves is a measure of the work performed by the tool in rearranging the silica network. These calculated loading and unloading curves are strikingly similar to those observed experimentally. It is essential to learn more about cracks for a better understanding of deformation in glass. To simulate the propagation of a crack into a silica surface, we put a notch on the top surface of the glass and let the computer simulation pull apart the sample at a constant velocity. The critical stress for crack propagation depends on the geometry and length of a pre-existing crack. Longer cracks, for example, yield at lower applied stress than do shorter cracks.

Appendices

APPENDIX - I

TOP 10 HOTTEST LABS IN NANOTECH

The most recent issue of Nanotechnology Law and Business (NLB, Vol. 1, Issue 4) provides a top ten list of "hottest" nanotechnology university labs. Foley & Lardner is actively working with four of the ten labs as intellectual property counsel, either directly with the university or through spin-out companies from these university labs. In its top ten list, NLB includes groups from California Institute of Technology (Professor Michael Roukes), Massachusetts Institute of Technology (Professor Angela Belcher), Northwestern University (Professor Chad Mirkin), and Rensselaer Polytechnic Institute (Professor Pulickel Ajayan). Foley & Lardner represents Nanotechnica, Cambrios, and NanoInk, which are spinout companies focused on commercially developing technology from the above described Caltech, MIT and Northwestern nanotechnology labs. NLB is a journal which explores the rapidly developing world of nanotechnology and miniaturization. Foley & Lardner attorneys Stephen Maebius, Leon Radomsky, and Steven Rutt handle work for these clients. For further information, contact Stephen Maebius at smaebius@foley.com or 202.672.5569.

To date, efforts to commercialize nanoscale materials, devices, and systems have largely been driven by academic pioneers. Research labs at different universities across the country have been responsible for some of the most promising and disruptive breakthroughs in nanotechnology. With a substantial federal commitment to funding nanoscience and the construction of new nanoscience centers at numerous universities, professors, undergrads, graduate students, and post-docs will spend an increasing amount of their time and energy on understanding nanoscale phenomenon and developing new technologies in the coming years. In this issue, we review the ten hottest university labs in nanotechnology. We identified research groups that have not just made great scientific strides, but are also actively leading the way toward commercialization of nanotechnology.

1. The Ajayan Group, Rensselaer Polytechnic Institute

The Ajayan group is using carbon nanotubes as templates and molds for fabricating nanowires, composites, and novel ceramic fibers. The group's research goals include producing macro-assemblies made of nanostructures for applications, understanding growth mechanisms of nanostructures and designing new structures and multifunctional nanocomposites. Other research interests are phase stability in metal clusters, the graphite-diamond phase transition, and growth of nanostructures under electron irradiation. Several start-ups have spun out of this group, and it collaborates with GE Global Research. The group is lead by Pulickel M. Ajayan, Professor of Materials Engineering at RPI.

2. The Alivisatos Group, UC Berkeley

Research in the Alivisatos group is primarily focused on nanocrystals. Alivisatos is Chancellor's Professor of Chemistry and Materials Science at UC Berkeley, Director of the Materials Sciences Division at the Lawrence Berkeley National Laboratory, and Director of the Molecular Foundry, a Department of Energy Nanoscale Science Research Center. Ongoing research projects in the Alivisatos group include nanocrystal / polymer composites for light emitting diodes and photovoltaics, and nanocrystal electron transistors, and nanocrystal/antibody conjugates as biological tag molecules, and DNA directed assembly of nanocrystal patterns. IP generated by this group relating to semiconductor nanomaterials is being developed at both Nanosys and its East Bay neighbor, Quantum Dot Corporation.

3. The Belcher Group, MIT

The Belcher group, which is comprised of 24 people, is exploring how to utilize biological systems to manufacture nanostructures. Angela Belcher's research, which dates back to her days as a graduate student at UCSB, involves modifying viruses to produce nanowires, self-assembling films, and other nanomaterials. Cambrios Technologies Corp., the company formed to commercialize her technology, is currently seeking to raise money and is a hot topic in venture capital circles. Belcher, only 37, has received numerous honors, including Small Times' 2002 innovator and researcher finalist awards; Beckman, DuPont and Army Young Investigator awards, and a Presidential Early Career Award. In September 2004, she was awarded a "genius grant" and the distinction of a MacArthur Fellow.

4. The Dai Group, Stanford University

The Dai group at Stanford University is likely to play a major role in the commercialization of carbon nanotubes. Dai discovered how to grow nanotubes in specific directions and orientations on substrates using a chemical vapor deposition process. His company, Molecular Nanosystems, holds intellectual property related to synthesis of nanotubes, nanotube sensors, and nanotubes as AFM tips. Nearly every academic and corporate research group working with carbon nanotubes has utilized Dai's techniques.

5. The Lieber Group, Harvard University

Charlie Lieber's goup is developing nanostructures for a variety of applications. The group's research is primarily focused on self-assembly techniques for axial nanowire structures, radial nanowire structures, and branched nanowire structures, with device applications such as biological / chemical sensing & detection, digital electronics & computing, and photonics. Nanosys has licensed a substantial portion of the IP portfolio from Lieber's lab. With funding from the Air Force, Army, Navy, DARPA, Ellison Medical Foundation, Intel Corporation, National Cancer Institute, and NSF, the Lieber group is likely to continue to make pioneering breakthroughs in the march toward commercialization of nanotechnology.

6. The Mirkin Group, Northwestern University

The Mirkin group focuses on developing methods for controlling the architecture of molecules and materials on the 1-100 nm length scale, and utilizing such structures in the development of analytical tools for chemical and biological sensing, lithography, catalysis, and optics. The Mirkin group pioneered the use of biomolecules as synthons in materials science and the development of nanoparticle-based biodiagnostics. Many of the concepts and materials developed within his laboratories are now the basis for commercial detection and lithography systems. Mirkin is the George B. Rathmann Professor of Chemistry and Director of the NU Institute for Nanotechnology. He is the inventor on over 50 patents, is an active consultant with several major chemical companies, and is the founder of two companies: Nanosphere and NanoInk.

7. The Quate Group, Stanford University

One of the most veteran research leaders on this list, Cal Quate has spent years developing imaging and lithography applications of scanning probes. He is a named inventor on 37 issued patents assigned to Stanford. Intellectual property developed by his group formed the basis for NanoDevices, a company that was acquired by Veeco. Quate's IP has also been licensed to Molecular Nanosystems, and he sits on the scientific advisory board of the company. The Quate group is currently focused on increasing throughput by scanning simultaneously with multiple probes all moving at high speeds. Specifically, the group's research involves the micromachining of cantilevers, fabrication of electron devices, system and circuit design, and the development and implementation of novel sensors and actuators.

8. The Roukes Group, Caltech

The Roukes group is developing and applying new techniques to create three-dimensional structures and machinery at the nanoscale. Roukes and his students have been pioneers in development of nanoelectromechanical systems. Roukes is the Director of Caltech's Kavli Nanoscience Institute, cofounder of the Nanosystems Biology Alliance, co-founder and co-director of both the Initiative in Computational Molecular Biology and the Laboratory for Large Scale Integration of Nanostructures, and chair of the external advisory board of Harvard University's nanoscience center. He recently teamed up with colleagues Yu-Chong Tai (Director of the Caltech Micromachining Laboratory) and Scott Fraser (Director of the Caltech Biological Imaging Center) to form Nanotechnica (division of Arrowhead Research Corporation), a company seeking to mass produce a variety of different nanoscale devices and systems.

9. The Smalley Group, Rice University

Perhaps the most important academic figure in the nanotech revolution is Richard Smalley, Professor of Chemistry and Professor of Physics & Astronomy at Rice University. Research in his lab lead to the discovery of the fullerene molecule and ultimately earned him a Nobel Prize in 1996. His current group is focused on single walled carbon nanotubes. The goal of the group is to develop methods of production, purification, derivitization, analysis, and assembly of nanotubes to solve real world problems. Carbon Nanotechnologies Inc., holder of the exclusive license to the suite of intellectual property developed in his lab, is developing a plant for mass production of single-walled carbon nanotubes.

10. The Whitesides Group, Harvard University

The Whitesides Group may be engaged in the most diverse and interdisciplinary areas of research of any group on this list, with projects stretching across biochemistry, materials science, catalysis, and physical organic chemistry. Ongoing projects include rational drug design, fabrication of nanostructures, microfluidic systems, soft lithography, fuel cells, and self-assembly of meso-scale systems. Perhaps the most vivid illustration of the impact that the Whitesides group could have on commercialization of nanotechnology is the role that graduate students played in guiding early efforts at Nanosys. Whitesides is the Mallinckrodt Professor of Chemistry at Harvard University and holds advisory positions with the National Science Foundation, National Research Council, and Department of Defense.

APPENDIX - II

NANO LABS & NANOTRIBOLOGY

The term tribology is derived from the Greek word *tribo* meaning rubbing and *logy* meaning knowledge. The original applications by the Greeks of tribology were in trying to understand the motion of large stones across the earth's surface. Today tribology has grown to include the methodical study of friction, lubrication, and wear. Tribology plays a critical role in diverse technological areas. In the advanced technological industries of semiconductor and data storage, tribological studies help optimize polishing processes and lubrication of data storage substrates. In traditional industries such as automotive and aerospace, tribological studies help increase the lifespan of mechanical components. Many industrial processes require a detailed understanding of tribology at the nanometer scale. The development of lubricants in the automobile industry depends on the adhesion of nanometer layers (mono layers) to a material surface. Assembly of components can depend critically on the adhesion of materials at the nanometer length scale.

There are a number of traditional tools for characterizing friction, lubrication and wear. The most common characterization tool is the tribometer having several configurations such as pin-on-disk, ball on flat, and flat on flat, etc. Generating motions at the nanometer scale is extremely challenging. New characterization techniques are required to understand tribology at the nanometer scale. The atomic force microscope is now being routinely applied for studying nanoscale tribology. The natural extension of the AFM for tribology applications is derived from the motion of a nanometer-sized stylus in the AFM over a surface. Although traditional tribology testing is not done with an AFM, many new types of applications are possible. Examples of the application of AFM to tribology include: (i) Direct three-dimensional visualization of wear tracks, or scars on a surface. (ii) Measurement of the thickness of solid and liquid lubricants having nanometer or even monolayer thickness.(iii) Measurement of frictional forces at the nanometer scale. (iv) Surface characterization of morphology, texture, and roughness. (v) Evaluation of mechanical properties such as hardness and elasticity, and plastic deformation at the nanometer scale. A major advantage of the AFM for tribological studies is that the AFM can be routinely used on all types of materials. Materials commonly studied include: ceramics, metals, polymers, semiconductors, magnetic, optical, and biomaterials. AFM investigations are usually made in ambient air environment. It is possible to make AFM studies in a vacuum or liquid environment.

APPENDIX - III

NANO LABS & MOLECULAR NANOTECHNOLOGY

Molecular nanotechnology ("MNT") is an anticipated manufacturing technology that would allow precise control and positional assembly of molecule-sized building blocks through the use of nano-scale manipulator arms. Molecular nanotechnology is usually considered distinct from the more inclusive term "nanotechnology", which is now used to refer to a wide range of scientific or technological projects that focus on phenomena or properties of the nanometer scale (around 0.1-100nm). Nanotechnology is already a blossoming field, but molecular nanotechnology - the goal of productive, molecular-scale machine systems - is still in the preliminary research stage.

Nanotechnology was first introduced in 1959, in a talk by the Nobel Prize-winning physicist Richard Feynman, entitled "There's Plenty of Room at the Bottom". Feynman proposed using a set of conventional-sized robot arms to construct a replica of themselves, but one-tenth the original size, then using that new set of arms to manufacture an even smaller set, and so on, until the molecular scale is reached. If we had many millions or billions of such molecular-scale arms, we could program them to work together to create macro-scale products built from individual molecules - a "bottom-up manufacturing" technique, as opposed to the usual technique of cutting away material until you have a completed component or product - "top-down manufacturing".

Feynman's idea remained largely undiscussed until the mid-80s, when the MIT-educated engineer K. Eric Drexler published "Engines of Creation", a book to popularize the potential of molecular nanotechnology. Because MNT would allow manufacturers to fabricate products from the bottom up with precise molecular control, a very wide range of chemically possible structures could be created. Since MNT systems could put every molecule in its specific place, molecular manufacturing processes could be very clean and efficient. Also, because every little bit of matter in an molecular nanotechnology system would be part of a nano-scale manipulator, nanotechnological systems could be far more productive and maintain much higher throughputs than modern manufacturing techniques, which use macro-scale manipulators to fabricate products.

To initiate an MNT revolution would require an "assembler" - a reprogrammable nano-scale manipulator capable of creating a wide range of molecular structures, including a complete copy of itself. The first assemblers will only function effectively in lab-controlled environments, such as a vacuum. The advent of self-replicating molecular nanomachines could quickly lead to "desktop nanofactories", tabletop appliances that consume modest amounts of power and contain the software required to manufacture an interesting range of useful products. The arrival of MNT would revolutionize wide sectors of human activity, including manufacturing, medicine, scientific research, communication, computing, and warfare. When full-blown molecular nanotechnology will arrive is currently unknown, but many experts foresee its arrival between 2010 and 2020.

APPENDIX - IV

FAQS ON NANOTECHNOLOGY & NANO LABS

What is Artificial Life?

Field of research that focuses on the processes of life and how to better understand them by simulating them in computers. Artificial life is an area of research that studies the processes of life (how life is created, becomes extinct, evolves, reproduces, etc.) and simulates them in computers to better understand them. Artificial life touches as much on computer science as biology. What is life? Vast question that computer scientists have taken their turn trying to answer by grabbing onto two key concepts: 1) life reproduces itself, and 2) life evolves. In the 1970s, an artificial life computer program travelled around the world: the 'game of life.' In it, ' cells ' (actually, black dots on the computer screen) appear, move, and die according to a set of simple rules. From an initial population distributed on the screen at random, stable structures emerge, some moving, some immobile, that resemble, in circumstance, what might have been the first living organisms. Today, artificial life calls on increasingly complex ideas, such as emergence, and touches more and more the fields of robotics and bionics

What are the Potential Dangers of Molecular Nanotechnology (MNT)?

If potential benefits of molecular nanotechnology (MNT) sound too good to be true, there is one caveat – the *potential dangers* of molecular nanotechnology. When nanofactories can arrange atoms into structures – playing with the building blocks of life itself, or in this case *nanoblocks* — theoretically anything allowable by the laws of physics can be created fast and cheap. Requirements include a few square feet for the nanofactory, the software, and an electrical outlet.

Criminals, terrorists, disturbed individuals, governments, and antisocial groups of all stripes would be incredibly empowered by such technology. Additional potential dangers of molecular nanotechnology threaten the economy, environment, human rights, and world peace. The rush to gain supremacy through nanoweaponry could lead to a new arms race, while attempts to stranglehold the technology would likely result in independent, covert development. Unilateral, "open-source" international cooperation is another option that runs its own risks, and control in the public sector could lead to inequitable benefits and an Orwellian society.

The probability factor of certain potential dangers of molecular nanotechnology will be higher than others, but all are possible within a scope of circumstances that, without prevention through forethought and planning, could feasibly come to pass. Some dangers cannot be discounted even with said planning, while others can reasonably be assumed to be goals of recognized subversive elements.

Concerted efforts are underway to devise the best course of action by anticipating these dangers in advance. The Center For Responsible Nanotechnology (CRN) works closely with experts in the field including K. Eric Drexler, the definitive authority on MNT. In brief, here is a sampling of some potential dangers of molecular nanotechnology:

NANOWEAPONRY: THE NEWARMS RACE - Nanofactories make the manufacture of many kinds of weapons possible with incredibly accurate computerized systems. While older technologies were both difficult and costly, nanoweapons could be manufactured easily and quickly. Conventional style weapons made more powerful and new weapons such as poison-carrying nanorobots could be made by the billions nearly cost-free and delivered remotely. Once inhaled, they might even be tailor-made to kill only people with specific genetic signatures, thus used as a means for ethnic cleansing. An arms race could trigger reckless development and testing of new weapons with unpredictable results. Experts agree this is probably the #1 potential danger of molecular nanotechnology.

ENVIRONMENTAL IMPACT AND EXISTENTIAL DANGERS - The use of nanofactories to make countless cheap, durable products could lead to 'disposable thinking' where products are created en mass and discarded in abundance, overwhelming recycling needs and the environment.

Poor nations might use biomass (carbon-rich trees) as fuel for nanofactories, leading to increased deforestation.

Experimentation in nano-augmentation of plants and animals (for example, to make them larger, smaller, faster, stronger, etcetera) could easily lead to runaway consequences in the wild ("green goo" vs "gray goo") that could threaten existing plants and animals, affect the food chain, and pose unforeseen threats to human life. This is a prime concern.

Ecophage ("gray goo"), though only a remote possibility due to the complexity of designing self-replicating nanorobots (*replibots*) capable of the task - and the heat signature the process would trigger that would alert watchdog systems in place – remains at least mentionable in that it is not impossible.

ECONOMIC IMPACT - Another high concern among the potential dangers of molecular nanotechnology is that many predict MNT will arrive suddenly and in full force. The sudden advent of nanofactories producing clean, cheap, durable, products would adversely impact most sectors in the job market. Skilled labor, factory workers, and many lines of distribution would no longer be needed as corporations switched to nanotechnology or folded. Stocks would be critically affected and the likelihood of economic upheaval, high.

DANGERS OF REGULATION - Though MNT has the potential to be the great equalizer, making products, medicine, and drinking water available to the entire world, its ubiquitousness would depend on how it is regulated, by whom, and to what purpose. Many corporations are likely to be motivated by potential windfall profits. They may legally protect then overprice nanotechnology, putting its benefits out of reach for those who need it most, while not passing the savings to the general public.

Other fronts of regulation also frame potential dangers of molecular nanotechnology. If development is too restrictive one set of problems is created (including inadvertent encouragement of a completely unregulated black market); and if restrictions are too lax, another set of problems is created (including possible damage to the environment and increased risk to the public).

UBIQUITOUS SURVEILLANCE - Again one of the benefits of MNT becomes a source for potential danger. Miniaturization of computer technology will allow unprecedented surreptitious surveillance of individuals. Spybots could be inhaled without even being aware. Increased computer power would allow a government to keep real time surveillance records on each and every citizen in a nation, no matter how large the population. The need to regulate the use of home nanofactories could conceivably be an excuse for such an invasion of privacy.

ARTIFICIAL INTELLIGENCE (AI) AND ROBOTICS - One of the most controversial dangers of molecular nanotechnology is that it will open the door to computers that think faster than the human brain, giving machines a superior edge. As robotics and AI combine to relieve humans of doing tasks that machines can do better, faster and cheaper, some believe we may be paving the way to our own destruction. Will nations secretly create armies of AI-enhanced, nano-augmented (think *bionic*) supersoldiers to fight wars? Will politicians opt for AI-enhancements? Nano-augmentation? Who will it be available to, and are we as a race headed towards total dependency on machinery to the extent it becomes part of our biology? Will there be equity or will a new class divide be created, similar to that depicted in *Gattica*? If we do not embrace AI-enhancement and nano-augmentation will intelligent machines ultimately decide we are unnecessary? We're not in Kansas anymore.

These suggested potential dangers represent a body of significant hurdles to overcome in the predawn stages of this very powerful technology, but one thing is clear: the cat can't be put back in the bag. If responsible nations ceased all development of MNT, this would not prevent other nations from developing it. Experts believe the best course of action is for responsible nations to be ahead of the curve, not just to establish a prudent course of development and international guidelines, but to anticipate and develop safeguards for a future that might require big solutions for microscopic problems with global or even existential consequences.

One thing seems sure: whether the potential dangers of molecular nanotechnology worry you or not, the next few decades are bound to be interesting.

What is Banoscience?

The study of phenomena and manipulation of materials at atomic, molecular and macromolecular scales, where properties differ significantly from those at a larger scale. Nanoscience is primarily the extension of existing sciences into the realms of the extremely small (*nanomaterials, nanochemistry, nanobio, nanophysics*, etc.) while *nanoengineering* represents the extension of the engineering fields into the nano- scale realm (*nanofabrication, nanodevices*, etc.). The exponential growth of nanoscience is largely due to the development of new instruments and related techniques that are used to "routinely" probe and manipulate material at the atomic and molecular level. *Scanning probe microscopies,* analytical electron- beam techniques, epitaxial growth facilities, and *synchrotron* radiation sources are all opening huge opportunities.

What is Avalanche Diode?

Silicon diode that is designed to break down and conduct at a specified reverse bias voltage. A common application is protecting electronic circuits against damaging high voltages. The avalanche diode is connected to the circuit so that it is reverse-biased. In other words, its cathode is positive with respect to its anode. In this configuration, the diode is non-conducting and does not interfere with the circuit. If the voltage increases beyond the design limit, the diode suffers avalanche breakdown, causing the harmful voltage to be conducted to earth. Avalanche breakdown is due to impact ionization. Under a small reverse bias, the diode is almost non-conducting, although a very small current still flows. When the reverse electric field across the p-n junction is large enough, the energy of the few electrons flowing is enough to ionize atoms in the silicon. The ionization process is self-reinforcing, causing a rapid increase in current (theoretically taking only picoseconds to reach its peak). Avalanche breakdown is not destructive, as long as the diode is not allowed to overheat. The Zener diode exhibits an apparently similar effect, but its operation is caused by a different mechanism, called Zener breakdown. Both effects are actually present in any such diode, but one may dominate the other. Diodes that conduct in the reverse direction when the reverse bias voltage exceeds the breakdown voltage. These are electrically very similar to Zener diodes, and are often mistakenly called Zener diodes, but break down by a different mechanism, the Avalanche Effect. This occurs when the reverse electric field across the p-n junction causes a wave of ionization, reminiscent of an avalanche, leading to a large current. Avalanche diodes are designed to break down at a well-defined reverse voltage without being destroyed. The difference between the avalanche diode (which has a reverse breakdown above about 6.2 V) and the Zener is that the channel length of the former exceeds the 'mean free path' of the electrons, so there are collisions between them on the way out. The only practical difference is that the two types have temperature coefficients of opposite polarities. Practical voltage reference circuits feature Zener and switching diodes connected in series and opposite directions to balance the temperature coefficient to near zero

What is MNT (Molecular Nanotechnology)?

Molecular nanotechnology ("MNT") is an anticipated manufacturing technology that would allow precise control and positional assembly of molecule-sized building blocks through the use of nano-scale manipulator arms. Molecular nanotechnology is usually considered distinct from the more inclusive term "nanotechnology", which is now used to refer to a wide range of scientific or technological projects that focus on phenomena or properties of the nanometer scale (around 0.1-100nm). Nanotechnology is already a blossoming field, but molecular nanotechnology - the goal of productive, molecular-scale machine systems - is still in the preliminary research stage.

Nanotechnology was first introduced in 1959, in a talk by the Nobel Prize-winning physicist Richard Feynman, entitled "There's Plenty of Room at the Bottom". Feynman proposed using a set of conventional-sized robot arms to construct a replica of themselves, but one-tenth the original size, then using that new set of arms to manufacture an even smaller set, and so on, until the molecular scale is reached. If we had many millions or billions of such molecular-scale arms, we could program them to work together to create macro-scale products built from individual molecules - a "bottom-up manufacturing" technique, as opposed to the usual technique of cutting away material until you have a completed component or product - "top-down manufacturing".

Feynman's idea remained largely undiscussed until the mid-80s, when the MIT-educated engineer K. Eric Drexler published "Engines of Creation", a book to popularize the potential of molecular nanotechnology. Because MNT would allow manufacturers to fabricate products from the bottom up with precise molecular control, a very wide range of chemically possible structures could be created. Since MNT systems could put every molecule in its specific place, molecular manufacturing processes could be very clean and efficient. Also, because every little bit of matter in an molecular nanotechnology system would be part of a nano-scale manipulator, nanotechnological systems could be far more productive and maintain much higher throughputs than modern manufacturing techniques, which use macro-scale manipulators to fabricate products.

To initiate an MNT revolution would require an "assembler" - a reprogrammable nano-scale manipulator capable of creating a wide range of molecular structures, including a complete copy of itself. The first assemblers will only function effectively in lab-controlled environments, such as a vacuum. The advent of self-replicating molecular nanomachines could quickly lead to "desktop nanofactories", tabletop appliances that consume modest amounts of power and contain the software required to manufacture an interesting range of useful products. The arrival of MNT would revolutionize wide sectors of human activity, including manufacturing, medicine, scientific research, communication, computing, and warfare. When full-blown molecular nanotechnology will arrive is currently unknown, but many experts foresee its arrival between 2010 and 2020.

What is Chaos Theory?

Chaos theory refers to the behavior of certain systems of motion, such as ocean currents or population growth, to be especially sensitive to tiny changes in starting conditions that result in drastically different outcomes. Unlike what it implies colloquially, chaos theory doesn't mean the world is metaphorically chaotic, nor does it refer to entropy, by which systems naturally tend toward disorder. Chaos theory relies on the uncertainty inherent in measurements, the precision of predictions, and the non-linear behavior of seemingly linear systems.

Before quantum mechanics, chaos theory was the first "weird" idea of physics. Back in 1900, Henri Poincaré thought about the relationship between values at different time points of a system whose general behavior could be accurately predicted, such as a planet in orbit. He realized that we can't ever exactly pinpoint a measurement, like position, speed, or time, because every instrument we could *possibly* develop would have a limit on its sensitivity. That is, no measurement is infinitely precise. Poincaré knew that motion is deterministically described by a series of equations that can accurately predict things like where a ball will end up if it is rolled down a ramp. However, he theorized that a tiny difference in initial conditions, based on almost insignificant variations in a measurement like mass, could result in two completely different macroscopic outcomes far, far in the future. This theory was called Dynamical Instability, and later scientists confirmed the veracity of his ideas.

Therefore, chaos theory studies how organized, stable systems cannot always yield meaningful predictions for a much later time, even though short-term behavior more closely follows expectations. In fact, any predictions it does yield might be so wildly divergent that they are no better than guesses. It is anti-intuitive that a more precise value would not yield a more precise output.

The snowball effect of a minute change in influential circumstances is referred to as the Butterfly Effect, whereby a butterfly flapping its wings (an almost imperceptible influence) could contribute to the development of a hurricane on the other side of the globe. Edward Lorenz did the first computer simulations in the 1960s that demonstrated dynamical instability with actual equations and data.

We cannot infer the initial conditions from later conditions, nor vice versa, in several important systems, such as atmospheric pressure and ocean currents that contribute to weather and climate. This is not merely a real-life scenario, resulting from something like too few thermometers in the ocean. Chaos theory is a verifiable, mathematically consistent theory that shows that sometimes increasingly precise measurements plugged into equations do not yield increasingly precise predictions, but rather such extreme diverging values that they are practically useless.

Some physicists are working on connections between this seeming randomness and large-scale structure, such as patterns in global climate, mass distribution of galaxies in superclusters, or population variation on a geologic time scale. They hypothesize that on a macroscopic level, certain kinds of organization and consistency have only been made possible through the disorder and inconsistency of chaos theory.

How Big is the Universe?

The current, observable universe has been determined to have a width of 156 billion light years, with an error of less than 1%, by the latest deep-space telescope WMAP. At first, it might seem impossible that scientists are so sure of this astronomical measurement, but this figure has been narrowed by years of research and determined by several paths of inquiry. Also, the size of the universe is intimately dependent on its shape, age, acceleration, and total mass, so we are very confident in this figure.

In 2003, the Wilkinson Microwave Anisotropy Probe sent back enough data for scientists to publish extremely dependable studies that established two previously unknown facts about the universe. They determined our universe is flat, which means standard Euclidean geometry is valid on the largest scale. This can be understood by saying a straight line more or less stays a straight line for as long as it extends. They also established that the universe is accelerating at an ever-increasing rate, which means that all mass is flying away from each other at faster and faster speeds. The WMAP data measured the temperature, called the cosmic microwave background radiation, of our observable universe with an unprecedented accuracy, to within a 5% error. From these facts, we can deduce figures such as the radius of the universe.

Remember that the size of the universe is not a constant value, nor is it the size of an object as we traditionally understand it. The size of the universe is actually the size of space itself, and as space expands, so does the space between planets, stars, and galaxies. At the beginning of the universe, the Big Bang created space and time as we know them. From that moment, space has been expanding, so we find its size by measuring how far light could have traveled since the Big Bang, along with how much space itself stretched.

We can only possibly look or communicate up to the edge, or "horizon," of where light has traveled since the beginning of the universe. The size of the universe means the space in which we can interact with anything. We will never ever know what is "beyond" this boundary, because there is no way to know anything about it, so it's illogical to consider the realm "outside" of our universe, or to wonder what we are expanding "into."

An independent measure of the size of our universe can be given by studying the oldest stars. The oldest stars we have found are probably somewhere between 11 and 14 billion light years old. If we had stars older than the largest distance light could have traveled, then we'd know there was something wrong with our calculations; there would not be enough time for them to evolve. However, these values are consistent with everything else we know about the universe.

What is a Neutrino?

A particle with very low mass (around that of an electron), and no electrical charge, the neutrino is an elusive subatomic particle. The neutrino is so shy that the duration between the theorization of its existence and its actual discovery was 25 years. Wolfgang Pauli, a famous quantum physicist, theorized the neutrino in 1931. It was discovered by Frederick Reines and Clyde Cowan in 1956 at a neutrino observatory located adjacent to a nuclear power plant at Savannah River, South Carolina.

Neutrinos travel at almost the speed of light, and many quadrillions of them penetrate your body every second. But because neutrinos have such low mass and interact only slightly with atoms, they can penetrate several light years of densely packed matter before interacting with an atom. For this reason they are very difficult to detect.

Neutrinos are generated during an event known in physics as beta decay. It seemed hopeless to detect neutrinos until the advent of nuclear technology. Atomic bombs and nuclear reactors proved to be rich sources of neutrino activity relative to a typical spot on Earth. The first neutrino detectors were tanks filled with water and cadmium chloride. The first neutrino detected was in fact not a conventional neutrino but an anti-neutrino.

When an anti-neutrino collided with a proton in the neutrino detector, the interaction produced a neutron and a positron, or an anti-electron. The resulting anti-electron would quickly annihilate with one of the electrons orbiting the nucleus, resulting in a spray of two photons. Then a stray neutron released from the breakdown of the atom would eventually (~15ms) be picked up by another, intact atom, releasing more photons (light). This distinct 2-stage pattern of photon release could be magnified by photoamplifiers, thereby triggering a register and providing positive evidence for neutrino impact.

With modern methods, as many as one neutrino per day is detected in our observatories. The neutrino is an excellent example of a fundamental particle that becomes more understandable as the quality of our scientific instruments improves. The continued gathering of evidence regarding the neutrino and its properties is sure to contribute in valuable ways to the progress of contemporary theoretical physics, which in turn will generate useful technological and theoretical discoveries for human civilization.

What is Carbon Nanofoam?

Carbon nanofoam is an allotrope of carbon. An allotrope is a variant of a substance composed of only one type of atom. The best-known allotropes of carbon are graphite and diamond. Carbon nanofoam, the 5th allotrope of carbon, was discovered in 1997 by Andrei V. Rode and his team at the Australian National University in Canberra, in collaboration with Ioffe Physico-Technical Institute in St Petersburg. Its molecular structure consists of carbon tendrils bonded together in a low-density, mistlike arrangement. Carbon nanofoam is similar in some respects to carbon and silicon aerogels produced before, but with about 100 times less density. Carbon nanofoam has been extensively studied under electron microscope by John Giapintzakis and team at the University of Crete. Its production and study has primarily been pioneered by Greek, Russian, and Australian scientists. The carbon nanofoam is produced by firing a high-pulse, high-energy laser at graphite or disordered solid carbon suspended in some inert gas such as argon. Like aerogels, carbon nanofoam has extremely high surface area and acts as a good insulator, capable of being exposed to thousands of degrees Fahrenheit before deforming. It is practically transparent in appearance, consisting of mostly air, and fairly brittle. One of the most unusual properties displayed by carbon nanofoam is that of ferromagnetism; it is attracted to magnets, like iron. This property vanishes a few hours after the nanofoam is made, though it can be preserved by cooling the nanofoam to extremely low temperatures, about -183° Celsius (-297° Fahrenheit). Other allotropes of carbon, such as fullerenes at high pressure, display some properties of magnetism, but not at the level carbon nanofoam does. The magnetic properties of carbon nanofoam remind scientists that the magnetism of a substance cannot be determined simply by the type of substance, but by its allotrope and temperature as well. (2) A new form of carbon: a spongy solid that is extremely lightweight and, unusually, attracted to magnets... John Giapintzakis of the University of Crete has used an electron microscope to study the structure of the nanofoam. He says it is the fifth form of carbon known after graphite, diamond and two recently discovered types: hollow spheres, known as buckminsterfullerenes or buckyballs, and nanotubes. Electrically conductive carbon nanofoams are a new material with many of the properties of traditional aerogel material. These materials are available in the form of monoliths, granules, powders and papers. They are synthetic, lightweight foams in which the solid matrix and pore spaces have nanometer- scale dimensions. Prepared by sol- gel methods, nanofoams typically have low density, continuous porosity, high surface area, and fine cell/ pore sizes. Carbon nanotubes are cylindrical structures made only from carbon atoms that are about 1 nm in diameter and 1-100 microns in length. In a graphite sheet, carbon atoms arrange themselves in a two-dimensional hexagonal lattice. Carbon nanotubes can be though of as a strip of graphite sheet that is rolled up to form a cylinder. There are many different ways to cut up a piece of graphite and roll it up to form a tube. The tubes can have different diameters and different chiralities. The chirality is the twist of the rows of atoms along the length of the tube. Sometimes the atom rows are parallel to the axis of the tube and sometimes the rows form a helix that winds along the tube. When one or more tubes grow inside another carbon nanotube, it is called a multiwalled nanotube. Carbon naotubes are very strong; they are 5 times as strong as steel for the same wieght. The electrical properties of carbon nanotubes depend on their diameter and their chirality. Some tubes are metallic and some are semiconductors. Research on carbon nanotubes at TU Delft, Nanotech Now's nanotube and buckyball page. A form of carbon related to fullerenes, except that the carbon atoms form extended hollow tubes instead of closed, hollow spheres. Carbon nanotubes can also form as a series of nested, concentric tubes. Carbon nanotubes can be used as nanometer-scale syringe needles for injecting molecules into cells and as nanoscale probes for making fine-scale measurements. Carbon nanotubes can be filled and capped, forming nanoscale test tubes or potential drug delivery devices. Carbon nanotubes can also be "doped," or modified with small amounts of other elements, giving them electrical properties that include fully insulating, semiconducting, and fully conducting. A new form of carbon: a spongy solid that is extremely lightweight and, unusually, attracted to magnets... John Giapintzakis of the University of Crete has used an electron microscope to study the structure of the nanofoam. He says it is the fifth form of carbon known after graphite, diamond and two recently discovered types: hollow spheres, known as buckminsterfullerenes or buckyballs, and nanotubes. Jim Giles, Scientists create fifth form of carbon, Nature 23 Mar. 2004. Electrically conductive carbon nanofoams are a new material with many of the properties of traditional aerogel material. These materials are available in the form of monoliths, granules, powders and papers. They are synthetic, lightweight foams in which the solid matrix and pore spaces have nanometer- scale dimensions. Prepared by sol- gel methods, nanofoams typically have low density, continuous porosity, high surface area, and fine cell/ pore sizes. The foams are also electrically conductive and have a high capacitance.

What is MEMS Applications?

Automotive: the first high volume application MEMS come of age

Imagine dust, floating in the air, gathering information on the weather, winds, humidity and pollution – all in a tiny cubic millimeter package. This is the "smart dust" scientists at UC Berkeley are developing in the laboratory today. Thousands or millions of these dust motes will sense the environment and communicate simultaneously using MEMS devices. Other applications under development range from such medical uses as hospital rooms that track patients and their medications and treatments, to the use of tiny surgical "scrapers" injected into arteries to clean out plaque, to military reconnaissance, to the detection of biological and chemical weapons.

Automotive: the first high volume application

After spending about 25 years in the lab, however, it was the automotive industry which first commercially embraced MEMS devices in the '90's as airbag accelerometers, recognizing the benefits of MEMS devices' small size, relative low cost and high degree of sensitivity.

Early airbags required the installation of several bulky accelerometers made of discrete components mounted in the front of the car, with separate electronics near the airbag (at a cost of over $50). Today, because of MEMS, the accelerometer and electronics are integrated on a single chip at a cost of under $10. The small size (about the dimensions of a sugar cube) provides a quicker response to rapid deceleration. And because of the very low cost, manufacturers are adding side impact airbags as well. The sensitivity of MEMS devices is also leading to improvements where size and weight of passengers will be calculated so the airbag response will be appropriate for each passenger.

There are many other automotive applications for MEMS either in use now or coming soon, including fuel pressure sensors, air flow sensors, tire pressure sensors with automatic built-in tire pumps, "smart" sensors for collision avoidance and skid detection, "smart" suspension for sport utility vehicles to reduce rollover risk, automatic seatbelt restraint and door locking, vehicle security, headlight leveling, and navigation.

MEMS come of age

Another wide deployment of MEMS is their use as micronozzles that direct the ink in inkjet printers. They are also used to create miniature robots (micro-robots) as well as micro-tweezers, and are used in video projection chips with a million moveable mirrors.

MEMS have been rigorously tested in harsh environments for defense and aerospace where they are used as navigational gyroscopes, sensors for border control and environmental monitoring, and munitions guidance. In medicine they are commonly used in disposable blood pressure transducers and weighing scales.

Out of these practical experiences have grown a raft of other applications which are quickly approaching everyday use. For example, engineers are utilizing the ability of MEMS devices to collectively assemble information and applying it to "smart roads." Smart roads would be covered with millions of MEMS sensors. The sensors would act as a blanket of information, gathering and transmitting data about road conditions to people charged with maintaining them. Problems with roadways would be detected and repaired before they became serious. These smart roads could also send information to cars equipped with Global Positioning Devices, informing the on-board computer of road hazards, accidents, and traffic. Another safety application includes the detection of black ice on roadways and the development of windshields with automatic glare resistance.

Other arenas into which MEMS are moving include package shipping, where MEMS shock sensors would rest inside packages to monitor time and any type of damage that may occur while the package is in transit. One commercial application that is coming to market within months is a compact skin analysis unit for medical and consumer use, including department store cosmetics counters and pharmacies. This tool will contain small sensor-based units to analyze the physical and chemical properties of skin surfaces.

An additional and significant growth area for MEMS today is in telecommunications where the technology is being used for *wireless applications* as well as in *optical networks*.

What is Field-Effect Transistor (FET)?

The schematic symbols for p- and n-channel MOSFETs. The symbols to the right include an extra terminal for the transistor body (allowing for a seldom-used channel bias) whereas in those to the left the body is implicitly connected to the source. The most common variety of field-effect transistors, the enhancement-mode MOSFET (metal-oxide semiconductor field-effect transistor) consists of a unipolar conduction channel and a metal gate separated from the main conduction channel by a thin layer of (SiO_2) glass. This is why an alternative name for the FET is 'unipolar transistor.' When a potential difference (of the proper polarity) is impressed across gate and source, charge carriers are introduced to the channel, making it conductive. The amount of this current can be modulated, or (nearly) completely turned off, by varying the gate potential. Because the gate is insulated, no DC current flows to or from the gate electrode. This lack of a gate current (as compared to the BJT's base current), and the ability of the MOSFET to act like a switch, allows particularly efficient digital circuits to be created, with very low power consumption at low frequencies. The power consumption increases markedly with frequency, because the capacitive loading of the FET control terminal takes more energy to slew at higher frequencies, in direct proportion to the frequency. Hence, MOSFETs have become the dominant technology used in computing hardware such as microprocessors and memory devices such as RAM. Bipolar transistors are more rugged and hence more useful for low-impedance loads and inductively reactive (e.g. motor) loads. Power MOSFETs become less conductive with increasing temperature and can therefore be applied in shunt, to increase current capacity, unlike the bipolar transistor, which has a negative temperature coefficient of resistance, and is therefore prone to thermal runaway. The downside of this is that, while the power FET can protect itself from overheating by diminishing the current through it, high temperatures need to be avoided by using a larger heat sink than for an equivalent bipolar device. Macroscopic FET power transistors are actually composed of many little transistors. They are stacked (on-chip) to increase breakdown potential and paralleled to reduce R_{on}, i.e. allowing for more current, bussing the gates to provide a single control (gate) terminal. The depletion mode FET is a little different. It uses a back-biased diode for the control terminal, which presents a capacitive load to the driving circuit in normal operation. With the gate tied to the source, a DFET is fully on. Changing the potential of a DFET (pulling an N-channel gate downward, for example) will turn it off, i.e. 'deplete' the channel (drain-source) of charge carriers. MOSFETs, formerly called IGFETs (for Insulated Gate Field-Effect Transistor) can be depletion-mode, enhancement-mode, or mixed-mode, but are almost always enhancement mode in modern commercial practice. This means that, with the source and gate tied together (thus equipotential) the channel will be off (high impedance or non-conducting). The n-channel device (reverse for P-channel), like in the DFET, is turned on by raising the potential of the gate. Typically, the gate on a MOSFET will withstand +-20V, relative to the source terminal. If one were to raise the gate potential of an n-channel device without limiting the current to a few milliamps, one would destroy the gate diode, like any other small diode. Why do we typically think of n-channel devices as the default? In silicon devices, the ones that use majority carriers that are electrons, rather than holes, are slightly faster and can carry more current than their P-type counterparts. The opposite is true in GaAs devices. Substituting the p-n-junction with a Schottky junction gives a MESFET (Metal Semiconductor Field Effect Transistor), used for GaAs and other III-V semiconductor materials. Using bandgap engineering in a ternary semiconductor, like AlGaAs gives a HEMT (High Electron Mobility Transistor). The FET is simpler in concept than the bipolar transistor and can be constructed from a wide range of materials. The most common use of MOSFET transistors today is the CMOS (complementary metallic oxide semiconductor) integrated circuit which is the basis for most digital electronic devices. These use a totem-pole arrangement where one transistor (either the pull-up or the pull-down) is on while the other is off. Hence, there is no DC drain, except during the transition from one state to the other, which is very short. As mentioned, the gates are capacitive, and the charging and discharging of the gates each time a transistor switches states is the primary cause of power drain. The C in CMOS stands for 'complementary.' The pull-up is a P-channel device (using holes for the mobile carrier of charge) and the pull-down is N-channel (electron carriers).

This allows busing of the control terminals, but limits the speed of the circuit to that of the slower P device (in silicon devices). The bipolar solutions to push-pull include 'cascode' using a current source for the load. Circuits that utilize both unipolar and bipolar transistors are called Bi-Fet. A recent development is called 'vertical P.' Formerly, BiFet chip users had to settle for relatively poor (horizontal) P-type FET devices. This is no longer the case and allows for quieter and faster analog circuits. A clever variant of the FET is the dual-gate device. This allows for two opportunities to turn the device off, as opposed to the dual-base (bipolar) transistor which presents two opportunities to turn the device on. FETs can switch signals of either polarity, if their amplitude is significantly less than the gate swing, as the devices (especially the parasitic diode-free DFET) are basically symmetrical. This means that FETs are the most suitable type for analog multiplexing. With this concept, one can construct a solid-state mixing board, for example. The power MOSFET has a 'parasitic diode' (back-biased) normally shunting the conduction channel that has half the current capacity of the conduction channel. Sometimes this is useful in driving dual-coil magnetic circuits (for spike protection), but in other cases it causes problems. The high impedance of the FET gate makes it rather vulnerable to electrostatic damage, though this is not usually a problem after the device has been installed. A more recent device for power control is the insulated-gate bipolar transistor, or IGBT. This has a control structure akin to a MOSFET coupled with a bipolar-like main conduction channel. These have become quite popular

What is Universal Assembler?

This is a term that was introduced by Eric Drexler. In his terminology, an assembler is enzyme or catalyst that brings two molecules together and connects them. DNA can be used to program the assembly of a wide variety of proteins. Drexler imagines a universal assembler to be something that could be programmed to fabricate anything. The following quote was taken from Eric Drexler's book Engines of creation in the section where he discusses universal assemblers. These second-generation nanomachines - built of more than just proteins - will do all that proteins can do, and more. In particular, some will serve as improved devices for assembling molecular structures. Able to tolerate acid or vacuum, freezing or baking, depending on design, enzyme-like second-generation machines will be able to use as "tools" almost any of the reactive molecules used by chemists - but they will wield them with the precision of programmed machines. They will be able to bond atoms together in virtually any stable pattern, adding a few at a time to the surface of a workpiece until a complex structure is complete. Think of such nanomachines as assemblers. Because assemblers will let us place atoms in almost any reasonable arrangement (as discussed in the Notes), they will let us build almost anything that the laws of nature allow to exist. In particular, they will let us build almost anything we can design - including more assemblers. The consequences of this will be profound, because our crude tools have let us explore only a small part of the range of possibilities that natural law permits. Assemblers will open a world of new technologies. Advances in the technologies of medicine, space, computation, and production - and warfare - all depend on our ability to arrange atoms. With assemblers, we will be able to remake our world or destroy it. So at this point it seems wise to step back and look at the prospect as clearly as we can, so we can be sure that assemblers and nanotechnology are not a mere futurological mirage. Uses raw atoms and molecules to construct consumer goods, and is pollution free. Can be programmed to build anything that is composed of atoms and consistent with the rules of chemical stability. Eric Drexler talks about these assemblers as nanorobots with telescoping manipulator arms that are capable of picking up individual atoms, and combining them however they are programmed.

What is Triac?

A triac is an electronic component equivalent to two silicon controlled rectifiers joined end to end or back to back with their gates connected together. This results in a two-directional electronic switch which can conduct current in both directions when it is triggered (turned on). It can be triggered by either a positive or a negative voltage. The device turns off again when the current through it drops below a certain threshold value, such as the end of a half-cycle of alternating current. This makes the triac a convenient switch for AC circuits. Applying a trigger pulse at a controllable point in an AC cycle allows one to control the percentage of current that flows through the triac. Lower power triacs are used in lots of applications, especially domestic electrical appliances, such as light dimme

What is UV Embossing?

Cold embossing uses UV light to polymerize and solidify materials after forming by the stamp. The starting materials are thin films of UV-curable materials. Typical materials may be monomers, or inorganic-organic hybrid polymers, which are polymerized by UV to polymers with desirable optical properties. The films may be deposited onto silicon, gallium arsenide, or similar wafers. The UV-transmissive patterned master template is pressed against the film, imprinting the pattern, which is UV cured before pressure is released. UV embossing allows accurately aligning and overlaying multiple levels of features. UV embossing has a clear advantage over hot embossing in allowing multi-level applications, since the first level is polymerized after exposure, making it irrevocably formed and dimensionally stable, regardless of subsequent layer applications and exposures. UV embossing is a low pressure, room temperature process, with little shrinkage of the pattern. It also allows partial UV casting, by limiting the illumination to maintain clear areas on the wafer. Cold embossing may be done on one or both sides of the wafer. Double-sided embossing allows micromolding.

What is Quantum Well?

A P-N-P junction in which the "N" layer is ~10 nm (where traditional physics leaves off and quantum effects take over) and an "electron trap" is created. "If one makes a heterostructure with sufficiently thin layers, quantum interference effects begin to appear prominently in the motion of the electrons. The simplest structure in which these may be observed is a *quantum well*, which simply consists of a thin layer of a narrower-gap semiconductor between thicker layers of a wider-gap material." See Center for Quantum Electronics U of Dallas.

What is Quantum Teleportation?

'Teleportation' (or copying) of the quantum state from one particle to another. Not really spectacular as in Star Trek serial where macroscopic objects (humans, weapons and similar) are 'teleported'. Present technology can 'teleport' quantum objects such as photons, electrons ans similar. It is important to note that the object itself is not teleported during this process. What actually happens is that the quantum state (e.g. spin number) is exactly copied from one object to another, both spatially and temporally present *prior* to 'teleportation'. The technique used is based on the Einstein-Podolsky-Rosen effect (EPR) or paradox, which relies on the so-called nonlocality of the wavefunction. An EPR source produces pair of entangled particles, one of which serves as a 'carbon copy'. The third particle (original, source particle to be copied, introduced besides the EPR pair) is 'entangled' with one of the particles from the EPR pair which enables the copying of its state to another EPR particle that has never interacted with the original, yet it becomes its copy. The experimental setup requires one measuring apparatus that entangles the original with one of the EPR particles and another apparatus that acts on another EPR particle which becomes the copy. The experimentalists operating the two apparatuses must be in contact i.e. the one that acts on the copy particle must know the results of the entanglement measurement on the source (original) particle and another EPR particle which is 'waisted' in the process. It should also be noted here that none of the experimentalists does not know the quantum state of the original particle, even when the whole process finishes, neither is it measured during the process.

What is Waste Recycling?

An advanced nanotechnology might be able to build filters that dynamically modify themselves to attract the contaminant molecules detected by the air and water quality sensors. Once attached to the filter, the filter could in principle move the offending molecules to a molecular laboratory for modifications to useful or at least inert products. A swarm might implement such an active filter if it was able to dynamically manufacture proteins that could bind contaminant molecules. The protein and bound contaminant might then be manipulated by the swarm for transportation. With a sufficiently advanced nanotechnology it might even be possible to directly generate food by non-biological means. Then agriculture waste in a self-sufficient space colony could be converted directly to useful nutrition. Making this food attractive will be a major challenge

What is Space Elevator?

Issacs and Pearson proposed a space elevator—a cable extending from the Earth's surface into space with a center of mass at geosynchronous altitude. If such a system could be built, it should be mechanically stable and vehicles could ascend and descend along the cable at almost any reasonable speed using electric power (actually generating power on the way down). The first incredibly difficult problem with building a space elevator is strength of materials. Maximum stress is at geosynchronous altitude so the cable must be thickest there and taper exponentially as it approaches Earth. Any potential material may be characterized by the taper factor—the ratio between the cable's radius at geosynchronous altitude and at the Earth's surface. For steel the taper factor is tens of thousands—clearly impossible. For diamond, the taper factor is 21.9 including a safety factor. Diamond is, however, brittle. Carbon nanotubes have a strength in tension similar to diamond, but bundles of these nanometer-scale radius tubes shouldn't propagate cracks nearly as well as the diamond tetrahedral lattice. Thus, if the considerable problems of developing a molecular nanotechnology capable of making nearly perfect carbon nanotube systems approximately 70,000 kilometers long can be overcome, the first serious problem of a transportation system capable of truly large scale transfers of mass to orbit can be solved. The next immense problem with space elevators is safety—how to avoid dropping thousands of kilometers of cable on Earth if the cable breaks. Active materials may help by monitoring and repairing small flaws in the cable and/or detecting a major failure and disassembling the cable into small elements

What is Wavelength Routing?

A concept developed by Cisco Systems where at each intermediate node between the end nodes, light coming in on one port at a given wavelength gets routed out of one and only one port. It allows carriers to leverage their core network technologies to support the massive growth of data services. It avoids wasting transmitted power, by channeling the energy transmitted by each node along a restricted path instead of letting it spread over the entire network.

What is Thermodynamical Stability?

Stability of a certain system with respect to changes in thermodynamical parameters. For nanoscopic systems, the most important parameter is temperature , i.e. stability of a system with respect to increase of temperature. Some of nanoscopic structures are stable only at sufficiently low temperatures, while at higher temperatures their structure tends to degrade. At higher temperatures, the atoms of which a nanostructure is built vibrate with larger amplitudes which may result in degradation of the structure. Additionally, nanostructured material, or a nanoparticle may be unstable when exposed to chemical influences from the surrounding, i.e. atoms and molecules. Some nanostructures are stable only in ultra-high vacuum conditions, which very much limits their use. Thermodynamical and chemical stability is a major concern in many applications involving nanoparticles and nanostructured materials.

What is X-ray Crystallography?

X-ray crystallography involves firing X-rays through the crystal of a molecule to produce a diffraction pattern. This pattern provides information on the structure of that crystal. For example, X-ray crystallography helped scientists discover that the DNA molecule exists as a double helix. For more information see Crystallography 101 (Lawrence Livermore National Laboratory, USA) and X-ray crystallography (The British Biophysical Society, UK).

What is Zephyr?

A common term used to describe lightweight fabrics. The Zephyr philosophy is *building compilers from parts*. Parts might include front ends, back ends, optimizers, and the glue that holds all these pieces together. You might even *generate parts automatically* from compact specifications. If you *describe your intermediate forms* using Zephyr's Abstract Syntax Description Language (ASDL), we can generate data-structure definitions in *C, C++, Java, Standard ML, and Haskell*. Your IR can be serialized on disk and *freely exchanged* among compiler passes written in these languages; asdlGen creates the glue that holds your compiler together. You can also *visualize your IR* using our graphical browser. To *compile code for multiple targets*, the state of the art is to use a code-generator generator, which lets you specify a translation from an intermediate code to a target instruction set. We're trying to separate the intermediate code from the instruction set, so we can develop *machine descriptions that can be reused*, not just with compilers, but also with other tools. The idea is to *generate the machine-dependent parts* from descriptions of instructions' semantics, of binary representations, or of other properties. Zephyr's Computer Systems Description Languages (CSDL) let you *describe as much or as little as you need* for your application. Getting high-quality code requires optimization. Zephyr's very portable optimizer (vpo) provides *instruction selection*, *instruction scheduling*, and classical *global optimization*. Vpo operates on programs represented as register-transfer lists (RTLs) in SSA form. All vpo optimizations are "plug and play," and it is easy to add new ones. Vpo provides a level playing field for *evaluating different optimizations*.

What is meant by Rosette nanotubes?

Rosette nanotubes are tubes of nanometric sizes obtained through the self-assembly of organic molecular components. The basic unit for the self-assembly of rosette nanotubes is a base borrowing the structural properties of guanine and cytosine. Thus, the basic unit is hydrophobic and has complementary donor-donor-acceptor and acceptor-acceptor-donor H-bond arrays on its sides. The self-assembly process takes place in two stages. At first, six basic units self-assemble into a supermacrocycle ring. In the second stage, supermacrocycle rings stack forming a nanotube with a central hollow pore. Although very recently discovered (2001) rosette nanotubes allow us to envision interesting applications due to their tubular structure combined with the flexibility given by organic chemistry, which opens up a large range of possibilities for their functionalization. It is thought that rosette nanotubes could one day be used for othopaedic implants since one can attach growth factors and specific bone recognition peptide sequences that will preferentially attract bone cell adhesion. Similarly to the peptide nanotubes, rosette nanotubes could act as antibiotics by inserting in to the membrane of bacteria and sucking up nutrients.

What are Fluorescent semiconductor nanocrystals?

Fluorescent semiconductor nanocrystals are nanometer sized pieces of crystal comprising from a few tens to several thousands or more of atoms. Due to a quantal effect, quantum confinement, their electronic and optical properties are size dependent. In particular, when shined with UV light they emit narrow spectrum fluorescence light whose "color" shifts towards shorter wavelengths as the size of the crystallites decreases. Nanocrystals can be synthesized dispersed in a solid matrix (SiO2, ZrO2,...), by sol-gel method, or they can be prepared as colloids by wet chemistry techniques. This latter method yields the best samples in terms of size homogeneity and purity of the fluorescence color. Applications making use of the nanocrystals take advantage of their robust and tunable electronic/optical properties. One rapidly growing application of nanocrystals is in the fields of biology and medicine. Several recent studies have shown the potential of nanocrystals for multicolor fluorescent labelling of live cells, in vivo imaging of growing organisms, single particle tracking in cell traficking, intracellular staining, etc. Other fields of application is as light absorbing or emitting centers in photovoltaic devices or electroluminescent diodes. Of interest also is using the nanocrystals as tags in microbeads to prepare molecular barcodes.

Fluorescence: fluorescence corresponds to the fast emission of a photon (quantum of light) by a molecule/particle following the absorption of an excitation photon, of higher energy. Fluorescence stops when the excitation is turned off, in contrast to phosphorescence. Molecules/particles which possess this property are fluorophores. The probability of this process is termed the quantum yield (maximum value 1).

What is Thermodynamical Stability?

Stability of a certain system with respect to changes in thermodynamical parameters. For nanoscopic systems, the most important parameter is temperature , i.e. stability of a system with respect to increase of temperature. Some of nanoscopic structures are stable only at sufficiently low temperatures, while at higher temperatures their structure tends to degrade. At higher temperatures, the atoms of which a nanostructure is built vibrate with larger amplitudes which may result in degradation of the structure. Additionally, nanostructured material, or a nanoparticle may be unstable when exposed to chemical influences from the surrounding, i.e. atoms and molecules. Some nanostructures are stable only in ultra-high vacuum conditions, which very much limits their use. Thermodynamical and chemical stability is a major concern in many applications involving nanoparticles and nanostructured materials.

What is Tensiometer?

Force measuring instrument for determining surface / interfacial tensions of liquids, and dynamic contact angles of (regular shaped) solids, fibers, powders and porous materials. The measurement of surface / interfacial tension of liquids and contact angles of liquids against solids as performed by a tensiometer is based on force measurements of the interaction of a solid probe with the interface of a liquid or between 2 immiscible liquids. In these experiments a probe is hung on a balance and brought into contact with the liquid interface. The forces experienced by the balance as the probe interacts with the surface of the liquid can be used to calculate surface / interfacial tension or contact angles. The forces present in this situation depend on the following factors; size and shape of the probe, contact angle of the liquid/solid interaction and surface tension of the liquid. The size and shape of the probe are easily controlled. For surface / interfacial tension measurements the contact angle of the liquid towards the probe is controlled to be zero (complete wetting). This is achieved by using probes with high energy surfaces. For example, probes made of a platinum/iridium alloy insure complete wetting and they are very easily cleaned in a reliable way. The mathematical interpretation of the force measurements depends on the shape of the probe used. Two types of probes are commonly used for surface / interfacial tension measurements, the Wilhelmy plate and DuNouy Ring. The Wilhelmy plate method utilizes the interaction of a platinum plate with the liquid interface being tested. The calculations for this technique are based on the geometry of a fully wetted plate in contact with, but not submerged in, the liquid. In this method the position of the probe relative to the surface is significant. Therefore the measurement is made at the so called zero depth of immersion of the probe, and the force acting on the probe at this position is registered and can directly be used to calculate the surface tension of the liquid when the perimeter of the plate is accurately known. If only limited quantity of the liquid to be tested is available one may consider using a thin round platinum rod as the probe. In such a case the measurement is exactly the same as with the Wilhelmy plate, but the probe dimensions of the probe will be smaller which affects the accuracy of the measurement and hence also might affect the reproducibility of the end results. The Du Nouy ring method utilizes the interaction of a platinum ring with the liquid interface being tested. The ring is submerged below the interface and subsequently raised upwards. As the ring moves upwards it raises a meniscus of the liquid. Eventually this meniscus tears from the ring and returns to it's original position. Prior to this event, the volume, and thus the force exerted, of the meniscus passes through a maximum value and begins to diminish prior to the actually tearing event. The calculation of surface or interfacial tension by this technique is based on the measurement of this maximum force. As an additional volume of liquid which is raised due to the proximity of one side of the ring to the other mathematical corrections are needed in order to obtain the correct surface / interfacial tension values. he tensiometric method for measuring dynamic contact angles (Advancing and Receding contact angles) measures the forces that are present when a sample of solid is brought into contact with a test liquid. If the forces of interaction, geometry of the solid and surface tension of the liquid are known the contact angle may be calculated. The solid sample has to have a symmetrical and regular shape (rod, cube, round rod, rectangle, wire etc.) in order to use this method. The advantages of this method are that the actual measurement is very simple to perform, one measurement gives an average for the whole sample and that wetting characteristics of single fibers can easily be obtained. he Tensiometer instrument also enables to determine the wettability characteristics i.e. contact angles of powders and porous solids by using the Washburn method (Powder Wettability Measurement).

What is Swarms?

Active materials can theoretically be made entirely of machines. These are sometimes called swarms since they consist of large numbers of identical simple machines that grasp and release each other and exchange power and information to achieve complex goals. Swarms change shape and exert force on their environment under software control. Although some physical prototypes have been built, at least one patent issued, and many simulations run, swarm potential capabilities are not well analyzed or understood. We briefly discuss some concepts here. Michael proposes brick-shaped machines of various sizes that slide past each other to assume a variety of shapes. He has generated a large number of videos showing computer simulations of simple motions. Although his web site contains rather extravagant claims, this work has received a U.K. patent. Yim built a small swarm with macroscopic (size in inches) components called polypod, built a simulator of polypod, and programmed it to move in various ways to study locomotion. There are two brick shaped components in polypod, one of which has two prismatic joints linked by a revolute joint. The second component is a cubic connector with no mechanical motion. Polypod is programmed by tables for each member of the swarm. Each member is programmed to move at various speeds in each degree of freedom for certain amounts of time. The swarm components are implicitly synchronized so there is no clock signal. Hall proposes a swarm with 10 micron dodecahedral components each with 12 arms that can move in and out, rotate a little, and grab and release each other. This concept is called the "utility fog." Hall estimates that the utility fog would have a density of 0.2, tensile strength of 1000 psi in action and 100,000 psi in a passive mode, and have a maximum shear rate of 100 km/second/meter. Bishop proposes a swarm consisting of 100 nanometer brick-shaped components that slide past each other to change shape. Globus proposes a swarm with two kinds of components — edges and nodes. The terms "node" and "edge" are chosen to correspond to those in graph theory. The roughly spherical nodes are capable of attaching to five edges (for a tetrahedral geometry with one free edge per node) and rotating each edge in pitch and yaw. The rod-like edges are capable of changing length, rotating around their long axis, and attaching/detaching to/from nodes.

What is Rebuilding Tissues?

Again, skin provides easy examples and may be a natural place to start in practice. People often want hair where they have bare skin, and bare skin where they have hair. Cell herding machines could move or destroy hair follicle cells to eliminate an unwanted hair, or grow more of the needed cells and arrange them into a working follicle where a hair is desired. By adjusting the size of the follicle and the properties of some of the cells, hairs could be made coarser, or finer, or straighter, or curlier. All these changes would involve no pain, toxic chemicals, or stench. Cell-herding devices could move down into the living layers of skin, removing unwanted cells, stimulating the growth of new cells, narrowing unnaturally prominent blood vessels, insuring good circulation by guiding the growth of any needed normal blood vessels, and moving cells and fibers around so as to eliminate even deep wrinkles. At the opposite end of the spectrum, cell herding will revolutionize treatment of life-threatening conditions. For example, the most common cause of heart disease is reduced or interrupted supply of blood to the heart muscle. In pumping oxygenated blood to the rest of the body, the heart diverts a portion for its own use though the coronary arteries. When these blood vessels become constricted, we speak of coronary-artery disease. When they are blocked, causing heart muscle tissue to die, we speak of someone "having a coronary," another term for heart attack. Devices working in the bloodstream could nibble away at atherosclerotic deposits, widening the affected blood vessels. Cell herding devices could restore artery walls and artery linings to health, by ensuring that the right cells and supporting structures are in the right places. This would prevent most heart attacks. But what if a heart attack has already destroyed muscle tissue, leaving the patient with a scarred, damaged, and poorly functioning heart? Once again, cell-herding devices could accomplish repairs, working their way into the scar tissue and removing it bit by bit, replacing it with fresh muscle fiber. If need be, this new fiber can be grown by applying a series of internal molecular stimuli to selected heart muscle cells to "remind" them of the instructions for growth that they used decades earlier during embryonic development. Cell-herding capabilities should also be able to deal with the various forms of arthritis. Where this is due to attacks from the body's own immune system, the cells producing the damaging antibodies can be identified and eliminated. Then a cell-herding system would work inside the joint where it would remove diseased tissues, calcified spurs, and so forth, then rework patterns of cells and intercellular material to form a healthy, smoothly working, and pain-free joint. Clearly, learning to repair hearts and learning to repair joints will have some basic technologies in common, but much of the research and development will have to be devoted to specific tissues and specific circumstances. A similar process—but again, specially adapted to the circumstances at hand—could be used to strengthen and reshape bone, correcting osteoporosis. In dentistry, this sort of process could be used to fill cavities, not with amalgam, but with natural dentin and enamel. Reversing the ravages of periodontal disease will someday be straightforward, with nanomedical devices to clean pockets, join tissues, and guide regrowth. Even missing teeth could be regrown, with enough control over cell behavior.

What is Quantum Mechanics?

A largely computational physical theory explaining the behavior of quantum phenomena, which incorporates the theory of special relativity. Despite dilignet attempts, general relativity has not been sucessfully incorporated into quantum mechanics. A physical model of chemical and optical phenomena, as well as the behaviour of matter in general on a small scale. Quantum mechanics describes a system of particles in terms of a wave function defined over the configuration space of the system. Although the concept of particles having distinct locations is implicit in the potential energy function that determines the wave function (e.g., of a ground-state system), the observable dynamics of the system cannot be described in terms of the motion of such particles from point to point. In describing the energies, distributions, and behaviors of electrons in nanometer-scale structures, quantum mechanical methods are necessary. Electron wave functions help determine the potential energy surface of a molecular system, which in turn is the basis for classical descriptions of molecular motion. Nanomechanical systems can almost always be described in terms of classical mechanics, with occasional quantum mechanical corrections applied within the framework of a classical model. A physical model of chemical and optical phenomena, as well as the behaviour of matter in general on a small scale. Quantum mechanics describes a system of particles in terms of a *wave function* defined over the *configuration space* of the system. Although the concept of particles having distinct locations is implicit in the *potential energy* function that determines the wave function (e.g., of a *ground-state* system), the observable dynamics of the system cannot be described in terms of the motion of such particles from point to point. In describing the energies, distributions, and behaviors of electrons in nanometer-scale structures, quantum mechanical methods are necessary. Electron *wave functions* help determine the *potential energy surface* of a molecular system, which in turn is the basis for *classical* descriptions of molecular motion. *Nanomechanical* systems can almost always be described in terms of classical mechanics, with occasional quantum mechanical corrections applied within the framework of a classical model.

What is Resonant Tunneling Devices (RTDs)?

Here we talk about the future nanoelectronic devices using quantum effects employed in semiconductor devices for fabrication of computer devices and high frequency oscillators. The underlying principles used for defining the perpendicular transport of electrons in low dimensional structures up to room temperature and its current and voltage characteristics are explained using energy band diagrams. MOSFETS with its special characteristic features serves two purposes. First case its acts as an excellent conductor device even with its scaling properties and even with small changes in its bias voltage results in large changes in its conductivity, hence its acts as an amplifier device. Though it functions as a two state device there some factors need to be considered for its implementation. The obstacles associated with MOSFET are high electric field, heat power dissipation, vanishing bulk properties, shrinkage of depletion layers and unevenness of the depletion layer. Corrections have to be applied for these devices. Once the devices approach the nanoelectric properties, the bulk properties of semiconductor devices can be replaced by quantum physics which depend on the electron properties, energy quantization effects and tunneling come into effect. When quantum effects are applied to semiconductor devices the consideration of doped materials no longer considered as major issue. These devices are of great importance for high speed electronics. Here we discuss the quantum properties, general used terms such a quantum dot, tunneling effect, island, subband and electron behavior in quantized energy levels to deal with the resonant tunneling devices and the application of the quantum properties in various devices such as Resonant Tunneling Diodes (RTDs), Resonant Tunneling Transistors (RTTs). Here we explain the theory related to tunneling and the mathematical derivation for transmission probability of the particle in the double barrier junction.

Bibliography

Attebery, B.: 2004, 'Dust, Lust, and Other Messages from the Quantum Wonderland', in: N.K. Hayles (ed.), *Nanoculture: Implications of the New Technoscience*, Bristol, UK: Intellect Books, pp. 161-172.

Baird, D.: 1993, 'Analytical Chemistry and the Big Scientific Instrumentation Revolution', *Annals of Science*, 50, 267-290.

Baird, D.; Nordmann, A. & Schummer, J. (eds.): 2004, *Discovering the Nanoscale*, Amsterdam: IOS Press.

Baker, R.T.K. Synthesis, properties and applications of graphite nanofibers. In *R&D status and trends*, ed. Siegel et al.

Ball, P.: 2002, 'Tutorial: Natural Strategies for the Molecular Engineer', *Nanotechnology*, 13, 15-28.

Bate, R., Frazier, G., Frensley, W., Reed., M., "An Overview of Nanoelectronics," *Texas Instruments Technical Journal*, July-August 1989, pp. 13-20.

Baum, R.; Drexler, K.E.; Smalley, R.: 2003, 'Nanotechnology: Drexler and Smalley Make the Case For and Against 'Molecular Assemblers', *Chemical and Engineering News*, 81, 37-42.

Beck, J.S., J.C. Vartuli, W.J. Roth, M.E. Leonowicz, C.T. Kresge, K.D. Schmitt, C.T.-W. Chu, D.H. Olsen, E.W. Shepard, S.B. McCullen, J.B. Higgins, and J.L. Schlenker. 1992. *J. Am. Chem. Soc.* 114:10834.

Bensaude-Vincent, B.: 2001, 'The Construction of a Discipline: Materials Science in the United States', *Historical Studies in the Physical and Biological Sciences*, 31, 223-248.

Bensaude-Vincent, B.: 2004, 'Two Cultures of Nanotechnology?', *Hyle: International Journal for Philosophy of Chemistry*, 10(2), 65-82.

Berube, D.M.: 2004, 'The Rhetoric of Nanotechnology', in: D. Baird, A. Nordmann & J. Schummer (eds.), *Discovering the Nanoscale*, Amsterdam: IOS Press, pp. 173-192.

Bijker, W.E.: 1995a, *Of Bicycles, Bakelite, and Bulbs: Toward a Theory of Sociotechnical Change*, MIT Press, Cambridge, MA.

Bijker, W.E.: 1995b, 'Sociohistorical Technology Studies', in: S. Jasanoff, G.E. Markle, J. Petersen, T. Pinch (eds.), *Handbook of Science and Technology Studies*, Sage, London, pp. 229-256.

Bijker, W.E.; Law, J.: 1992, 'Do Technologies Have Trajectories?', in: W.E. Bijker, J. Law (eds.), *Shaping Technology/Building Society: Studies in Sociotechnical Change*, MIT Press, Cambridge, MA, pp. 1-16.

Bijker, W.E.; Pinch, T.: 1987, 'The Social Construction of Facts and Artifacts: Or How the Sociology of Science and the Sociology of Technology Might Benefit Each Other', in: W.E. Bijker, T.P. Hughes, T. Pinch (eds.), *The Social Construction of Technological Systems: New Directions in the Sociology and History of Technology*, MIT Press, Cambridge, MA, pp. 17-50.

Bimber, B.: 1994, 'Three Faces of Technological Determinism', in: M.R. Smith, L. Marx (eds.), *Does Technology Drive History?: The Dilemma of Technological Determinism*, MIT Press, Cambridge, MA, pp. 79-100.

Bowes, C.L., A. Malek, and G.A. Ozin. 1996. *Chem. Vap. Deposition* 2:97.

Braun, P.V., P. Osenar, and S.I. Stupp. 1996. *Nature*. 368: 2.

Braun, T.; Schubert, A. & Zsindely, S.: 1997, 'Nanoscience and nanotechnology on the balance', *Scientometrics*, 38, 321-325.

Brinker, C.J. 1996. *Curr. Opin. Solid State Mater. Sci.* 1:798.

Brooks, L.J.: 2003, *The Institute for the Future: A Site for Mapping, Maintaining, and Transforming Images of the Future*, (Ph.D. dissertation) University of California, San Diego, La Jolla.

Brown, N.: 2004, 'Needle on the Real: Technoscience and Poetry at the Limits of Fabrication', in: N.K. Hayles (ed.), *Nanoculture: Implications of the New Technoscience*, Bristol, UK: Intellect Books, pp. 173-190.

Bueno, O.: 2004, 'The Drexler-Smalley Debated on Nanotechnology: Incommensurability at Work?', *Hyle: International Journal for Philosophy of Chemistry*, 10(2), 83-98.

Bueno, O.: 2004, 'Von Neumann, Self-Reproduction and the Constitution of Nanophenomena', in: D. Baird, A. Nordmann & J. Schummer (eds.), *Discovering the Nanoscale*, Amsterdam: IOS Press, pp. 101-115.

Bush, V.: 1945, *Science, the Endless Frontier*, US Government Printing Office, Washington, DC.

Carroll, J.S.: 2001, 'Social Science Research Methods for Assessing Societal Implications of Nanotechnology', in: M.C. Roco and W.S. Bainbridge (eds.), *Societal Implications of Nanoscience and Nanotechnology*, Dordrecht: Kluwer, pp. 188-192.

Chen, C-Y., S.L. Burkett, H.-X. Li, and M.E. Davis. 1993. *Microporous Mater*. 2:27.

Chianelli, R.R. 1998. Synthesis, fundamental properties and applications of nanocrystals, sheets, and fullerenes based on layered transition metal chalcogenides. In *R&D status and trends,* ed. Siegel et al.

Chianelli, R.R., M. Daage, and M.J. Ledoux. 1994. *Advances in Catalysis* 40:177.

Claeson, T., and Likharev, K.," Single Electronics," *Scientific American,* June 1992, pp. 80-85.

Clausen-Schaumann, H.; Seitz, M.; Krautbauer, R.; Gaub, H.E.: 2000, 'Force Spectroscopy with Single Bio-Molecules', *Current Opinion in Chemical Biology*, 4, 524-530.

Coenen, C.: 2004, 'Nanofuturismus: Anmerkungen zu seiner Relevanz, Analyse und Bewertung', *Technikfolgenabschätzung ' Theorie und Praxis*, 13 (2), 78-85.

Constant, E.W., II: 1980, *The Origins of the Turbojet Revolution*, Johns Hopkins University Press, Baltimore.

Crandall, B.: 1999, 'Molecular Engineering', in: B. Crandall (ed.), *Nanotechnology: Molecular Speculations on Global Abundance*, MIT Press, Cambridge, MA, pp. 1-46.

Crandall, B.C. and Lewis, J., *Nanotechnology: Research and Perspectives*, MIT Press, Cambridge, 1992.

Crow, M. & Sarewitz, D.: 2001, 'Nanotechnology and Societal Transformation', in: M.C. Roco and W.S. Bainbridge (eds.), *Societal Implications of Nanoscience and Nanotechnology*, Dordrecht: Kluwer, pp. 45-54.

Decker, M.; Fiedeler, U.; Fleischer, Th.: 2004, 'Ich sehe was, was Du nicht siehst... zur Definition von Nanotechnologie', *Technikfolgenabschätzung ' Theorie und Praxis*, 13 (2), 10-16.

Dresselhaus, M., and G. Dresselhaus. 1995. *Ann. Rev. Mat. Sci*. 25:487.

Dresselhaus, M.S. 1998. Carbon-based nanostructures. In *R&D status and trends*, ed. Siegel et al.

Dresselhaus, M.S., G. Dresselhaus, and P. Eklund. 1996. *Science of fullerenes and carbon nanotubes*. San Diego: Academic Press.

Drexler, K.E.: 1981, 'Molecular Engineering – an Approach to the Development of General Capabilities for Molecular Manipulation', *Proceedings of the National Academy of Sciences of the United States of America-Physical Sciences*, 78, 5275-5278.

Drexler, K.E.: 1992, *Nanosystems: Molecular Machinery, Manufacturing, and Computation*, Wiley, New York.

Drexler, K.E.: 2004, 'Nanotechnology: From Feynman to Funding', *Bulletin of Science, Technology & Society*, 24, 21-27.

Dupuy, J.-P.: 2004, 'Complexity and Uncertainty: A Prudential Approach To Nanotechnology', in: European Commission (Community Health and Consumer Protection): *Nanotechnologies: A Preliminary Risk Analysis on the Basis of a Workshop, Brussels,* 1-2 March 2004, pp. 71-93.

Dupuy, J.P.: 2004, 'Pour une évaluation normative du programme nanotechnologique', *Annales des Mines*, (February 2004), 27-32.

Eigler, D.M.; Schweizer, E.K.: 1990, 'Positioning Single Atoms With a Scanning Tunneling Microscope', *Nature*, 344, 524-526.

Einsiedel, E.F.; Goldenberg, L.: 2004, 'Dwarfing the Social? Nanotechnology Lessons from the Biotechnology Front', *Bulletin of Science, Technology & Society*, 24, 28-33.

ETC: 2003a, *The Big Down: Atom Tech ' Technologies Converging at the Atomic Scale*. Winnipeg, Canada: Action Group on Erosion, Technology and Concentration [www.etcgroup.org/documents/TheBigDown.pdf].

ETC: 2003b, 'No Small Matter II: The Case for a Global Moratorium ' Size Matters!' *Occasional Paper Series* 7(1) [www.etcgroup.org/documents/Occ.Paper_Nanosafety.pdf].

Etzkowitz, H., 2001, 'Nano-Science and Society: Finding a Social Basis for Science Policy', in: M.C. Roco and W.S. Bainbridge (eds.), *Societal Implications of Nanoscience and Nanotechnology*, Dordrecht: Kluwer, pp. 121-128.

European Comission: *European Workshop on Social and Economic Research on Nanotechnologies and Nanosciences, Brussels, 14-15 April 2004* [www.stage-research.net/STAGE/PAGES/Nano.html].

European Commission (Community Health and Consumer Protection): 2004, *Nanotechnologies: A Preliminary Risk Analysis on theBasis of a Workshop, Brussels, 1-2 March 2004.*

European Commission: 2004, *Converging Technologies: Shaping the Future of European Societies (Report of the High Level Expert Group "Foresighting the New Technology Wave")*, Brussels: European Commission Research [http://europa.eu.int/comm/research/conferences/2004/ntw/index_en.html].

F. Buot, "Mesoscopic Physics and Nanoelectronics: Nanoscience and Nanotechnology," *Physics Reports*, pp.73-174, 1993.

F. Capasso and S. Datta, "Quantum Electron Devices," *Physics Today*, pp.74-82, February 1990.

Feynman, R., "There's Plenty of Room at the Bottom: An invitation to Enter a New Field of Physics," Talk at the Annual Meeting of the American Physical Society, 29 December 1959.

Feynman, R.: 1999 [1959], 'There's Plenty of Room at the Bottom: An Invitation to Enter a New Field of Physics', in: *The Pleasure of Finding Things Out*, Perseus, Cambridge, MA, pp. 117-139.

Fielder, F.A.; Reynolds, G.H.: 1994, 'Legal Problems of Nanotechnology: An Overview', *Southern California Interdisciplinary Law Journal*, 3, 593-629.

Fleischer, Th.; Decker, M. & Fiedeler, U. (eds.): 2004, *Große Aufmerksamkeit für kleine Welten ' Nanotechnologie und ihre Folgen*, special issue of *Technikfolgenabschätzung ' Theorie und Praxis*, 13 (2), 5-85 [www.itas.fzk.de/tatup/042/inhalt.htm].

Fogelberg, H. & Glimell, H.: 2003, 'Molecular Matters: In Search of the Real Stuff', in: H. Fogelberg & H. Glimell, *Bringing Visibility to the Invisible: Towards A Social Understanding of Nanotechnology*, Göteborg: Göteborg University, pp. 5-32 [www.sts.gu.se/publications/STS_report_6.pdf].

Fogelberg, H.: 2003, 'The Material Culture of Nanotechnology', in: H. Fogelberg & H. Glimell, *Bringing Visibility to the Invisible: Towards A Social Understanding of Nanotechnology*, Göteborg: Göteborg University, pp. 99-114.

Francoeur, E.: 1997, 'The Forgotten Tool: The Design and Use of Molecular Models', *Social Studies of Science*, 27, 7-40.

Frazier, G., "An Ideology For Nanoelectronics," in *Concurrent Computations: Algorithms, Architecture, and Technology*, Plenum Press, New York, 1988.

Galison, P.: 1996, 'Computer Simulations and the Trading Zone', in: P. Galison, D.J. Stump (eds.), *The Disunity of Science: Boundaries, Contexts, and Power*, Stanford University Press, Stanford, pp. 118-157.

Galison, P.: 1997, *Image and Logic: A Material Culture of Microphysics*, University of Chicago Press, Chicago.

Galison, P.; Stump, D.J. (eds.): 1996, *The Disunity of Science: Boundaries, Contexts, and Power*, Stanford University Press, Stanford.

Gieryn, T.F.: 1999, *Cultural Boundaries of Science: Credibility on the Line*, University of Chicago Press, Chicago.

Gieryn, T.F.; Figert, A.E.: 1986, 'Scientists Protect Their Cognitive Authority: The Status Degradation Ceremony of Sir Cyril Burt', in: G. Böhme, N. Stehr (eds.), *The Knowledge Society*, Reidel, Dordrecht, pp. 67-89.

Glimell, H.: 2001, 'Challenging Limits' Excerpts from an Emerging Ethnography of Nano Physicists', in: H. Glimell & O. Johlin (eds.), *The Social Production of*

Technology: On the Everyday Life with Things, Göteburg, SE: BAS Publisher, chapter 7, pp. 111-131 (reprinted in: H. Fogelberg & H. Glimell, *Bringing Visibility to the Invisible: Towards A Social Understanding of Nanotechnology*, Göteborg: Göteborg University, pp. 115-139.

Glimell, H.: 2001, 'Dynamics of the Emerging Field of Nanoscience', in: M.C. Roco and W.S. Bainbridge (eds.), *Societal Implications of Nanoscience and Nanotechnology*, Dordrecht: Kluwer, pp. 156-160.

Glimell, H.: 2003, 'Dynamics of the Emerging Field of Nanoscience', in: H. Fogelberg & H. Glimell, *Bringing Visibility to the Invisible: Towards A Social Understanding of Nanotechnology*, Göteborg: Göteborg University, pp. 79-85 [www.sts.gu.se/publications/STS_report_6.pdf].

Glimell, H.: 2004, 'Grand Visions and Lilliput Politics: Staging the Exploration of the 'Endless Frontier'', in: D. Baird, A. Nordmann & J. Schummer (eds.), *Discovering the Nanoscale*, Amsterdam: IOS Press, pp. 231-246.

Goldhaber-Gordon, D., Montemerlo, M.S., Love, J.C., Opiteck, G.J., and Ellenbogen, J.C., "Overview of Nanoelectronic Devices," submitted to the *Proceedings of the IEEE*, February 1997. For more information, please send e-mail to nanotech@mitre.org.

Gorman, M.E.; Groves, J.F. & Shrager, J.: 2004, 'Societal Dimensions of Nanotechnology as a Trading Zone: Results from a Pilot Project', in: D. Baird, A. Nordmann & J. Schummer (eds.), *Discovering the Nanoscale*, Amsterdam: IOS Press, pp. 63-73.

Grinbaum, A. & Dupuy, J.-P.: 2004, 'Living with Uncertainty: Toward the Ongoing Normative Assessment of Nanotechnology', *Techne: Research in Philosophy and Technology*, 8(3) (forthcoming).

Grinbaum, A.: 2004, 'La condition de l'homme moderne et les nanotechnologies', in: G. Nivat (ed.), *Les limites de l'humain. 39èmes Rencontres Internationales de Genève*, L'Age d'Homme, Genève, p. 141.

Gross, M.: 1999, *Travels to the Nanoworld: Miniature Machinery in Nature and Technology*, Plenum, New York.

Grunwald, A.: 2004, 'Ethische Aspekte der Nanotechnologie. Eine Felderkundung', *Technikfolgenabschätzung ' Theorie und Praxis*, 13 (2), 71-78.

Gupta, V.K. & Pangannaya, N.B.: 2000, 'Carbon nanotubes: bibliometric analysis of patents', *World Patent Information*, 22, 185-189.

Guthold, M.; Falvo, M.R.; Matthews, W.G.; Paulson, S.; Washburn, S.; Erie, D.A.; Superfine, R.; Brooks, F.P.; Taylor, R.M.: 2000, 'Controlled Manipulation of Molecular Samples with the NanoManipulator', *IEEE-ASME Transactions on Mechatronics*, 5, 189-198.

Hacking, I.: 1983, *Representing and Intervening: Introductory Topics in the Philosophy of Natural Science*, Cambridge University Press, Cambridge, UK.

Hacking, I.: 1992, 'The Self-Vindication of the Laboratory Sciences', in: A. Pickering (ed.), *Science as Practice and Culture*, University of Chicago Press, Chicago, pp. 29-64.

Hansson, S.O.: 2004, 'Great Uncertainty about Small Things', *Techne: Research in Philosophy and Technology*, 8(3) (forthcoming).

Haruta, M. 1997. *Catalyst surveys of Japan* 1:61 and references therein.

Hayles, N.K. (ed.): 2004, *Nanoculture: Implications of the New Technoscience*, Bristol, UK: Intellect Books.

Hayles, N.K.: 2004, 'Connecting the Quantum Dots: Nanotechscience and Culture', in: N.K. Hayles (ed.), *Nanoculture: Implications of the New Technoscience*, Bristol, UK: Intellect Books, pp. 11-26.

Hecht, G., Allen, M.T.: 2001, 'Introduction: Authority, Political Machines, and Technology's History', in: M.T. Allen, G. Hecht (eds.), *Technologies of Power: Essays in Honor of Thomas Parke Hughes and Agatha Chipley Hughes*, MIT Press, Cambridge, MA, pp. 1-23.

Hecht, G.: 1998, *The Radiance of France: Technology, Politics, and National Identity after World War II*, MIT Press, Cambridge, MA.

Heidegger, M.: 1962, *Being and Time*, trans. J. Macquarrie, E. Robinson, Harper & Row, New York.

Heidegger, M.: 1977, *The Question Concerning Technology and Other Essays*, Harper Colophon Books, New York.

Helmreich, S.: 1998, *Silicon Second Nature: Culturing Artificial Life in a Digital World*, University of California Press, Berkeley, CA.

Hemerén, G.: 2004, 'Nano Ethics Primer', in: European Commission (Community Health and Consumer Protection): *Nanotechnologies: A Preliminary Risk Analysis on theBasis of a Workshop, Brussels, 1-2 March 2004* (www.europa.eu.int/comm/health/ ph_risk/documents/ev_20040301_en.pdf), pp. 95-102.

Hennig, J.: 2004, 'Changes in the Design of Scanning Tunneling Microscopic Images from 1980 to 1990', *Techne: Research in Philosophy and Technology*, 8(3) (forthcoming).

Hessenbruch, A.: 2004, 'Nanotechnology and the Negotiation of Novelty', in: D. Baird, A. Nordmann & J. Schummer (eds.), *Discovering the Nanoscale*, Amsterdam: IOS Press, pp. 135-144.

Hughes, T.P.: 1983, *Networks of Power: Electrification in Western Society, 1880-1930*, Johns Hopkins University Press, Baltimore.

Hughes, T.P.: 1994, 'Technological Momentum', in: M.R. Smith, L. Marx (eds.), *Does Technology Drive History? The Dilemma of Technological Determinism*, MIT Press, Cambridge, MA, pp. 101-114.

Hullmann, A. & Meyer, M.: 2003, 'Publications and Patents in Nanotechnology. An overview of previous studies and the state of the art', *Scientometrics*, 58 (3) 507-527.

Huo, Q., D.I. Margolese, U. Ciesla, P. Feng, T.E. Gier, P. Sieger, R. Leon, P.M. Petroff, F. Schuth, and G.D. Stucky. 1994. *Nature* 368:317.

Ihde, D.: 1991, *Instrumental Realism: The Interface between Philosophy of Science and Philosophy of Technology*, Indiana University Press, Bloomington.

Jena, P., S.N. Khanna, and B.K. Rao. 1996. In *Science and technology of atomically engineered materials*, ed. P. Jena. River Edge, NJ: World Scientific.

Johansson, M.: 2003, 'Plenty of room at the bottom: Towards an anthropology of nanoscience', *Anthropology Today*, 19 (No. 6), 3-6.

Johnson, A.: 2004, 'The End of Pure Science: Science Policy from Bayh-Dole to the NNI', in: D. Baird, A. Nordmann & J. Schummer (eds.), *Discovering the Nanoscale*, Amsterdam: IOS Press, pp. 217-230.

Jounet, C., W.K. Maser, P. Bernier, A. Loiseau, M. Lamy de la Chapelle, S. Lefrant, P. Deniard, R. Lee, and J.E. Fischer. 1997. *Nature* 388:756.

K.E. Drexler, *Nanosystems: Molecular Machinery, Manufacturing, and Computation*, Wiley, New York, 1992.

Karger, J., and D.M. Ruthven. 1992. *Diffusion in zeolites*. New York: J. Wiley.

Kay, L.: 2000, *Who Wrote the Book of Life? A History of the Genetic Code*, Stanford University Press, Stanford.

Keiper, A.: 2003, 'The Nanotechnology Revolution', *The New Atlantis*, 2, 17-34.

Khushf, G.: 2004, 'A Hierarchical Architecture for Nano-scale Science and Technology: Taking Stock of the Claims About Science Made By Advocates of NBIC Convergence', in: D. Baird, A. Nordmann & J. Schummer (eds.), *Discovering the Nanoscale*, Amsterdam: IOS Press, pp. 21-33.

Khushf, G.: 2004, 'Systems Theory and the Ethics of Human Enhancement: A Framework for NBIC Convergence', *Annals of the New York Academy of Sciences*, 1013, 124-149.

Kline, R.: 1992, *Steinmetz: Engineer and Socialist*, Johns Hopkins University Press, Baltimore.

Kline, R.: 1995, 'Construing 'Technology' as 'Applied Science": Public Rhetoric of Scientists and Engineers in the United States, *Isis*, 86, 194-221.

Kline, R.: 2000, 'The Paradox of 'Engineering Science': A Cold War Debate about Education in the United States', *IEEE Technology and Society Magazine*, 19, 19-25.

Kline, R.; Pinch, T.: 1996, 'Users as Agents of Technological Change: The Social Construction of the Automobile in the Rural United States', *Technology and Culture*, 37, 763-95.

Knorr-Cetina, K.: 1992, 'The Couch, the Cathedral, and the Laboratory: On the Relationship between Experiment and Laboratory in Science', in: A. Pickering (ed.), *Science as Practice and Culture*, University of Chicago Press, Chicago, pp. 113-138.

Knorr-Cetina, K.: 1999, *Epistemic Cultures: How the Sciences Make Knowledge*, Harvard University Press, Cambridge, MA.

Kranakis, E.: 1997, *Constructing a Bridge: An Exploration of Engineering Culture, Design, and Research in Nineteenth-Century France and America*, MIT Press, Cambridge, MA.

Krätschmer, W., L.D. Lamb, K. Fostiropoulos, and D.R. Huffman. 1990. *Nature* 347:354.

Kresge, C.T., M.E. Leonowicz, W.J. Roth, J.C. Vartuli, and J.S. Beck. 1992. *Nature* 359:710.

Kuhn, T.S.: 1996, *The Structure of Scientific Revolutions*, University of Chicago Press, Chicago.

Kupperman, A., S. Nadimi, S. Oliver, G. Ozin, J. Garcés, and M. Olken. 1993. *Nature* 365:239.

Kuusi, O.; Meyer, M.: 2002, 'Technological generalizations and leitbilder' the anticipation of technological opportunities', *Technological Forecasting & Social Change*, 69, 625-639.

Landon, B.: 2004, 'Less is More: Much Less is Much More: The Insistent Allure of Nanotechnology Narratives in Science Fiction', in: N.K. Hayles (ed.), *Nanoculture: Implications of the New Technoscience*, Bristol, UK: Intellect Books, pp. 131-146.

Laszlo, P.: 2004, 'Is There Life After Partington?', *Hyle: International Journal for Philosophy of Chemistry*, 10(2), 169-178.

Latour, B.: 1983, 'Give Me a Laboratory and I Will Raise the World', in: K. Knorr-Cetina, M. Mulkay (eds.), *Science Observed*, Sage, London, pp. 141-170.

Latour, B.: 1996, *Aramis or The Love of Technology*, Harvard University Press, Cambridge, MA.

Layton, E.T., Jr.: 1971, 'Mirror-Image Twins: The Communities of Science and Technology in 19th-Century America', *Technology and Culture*, 12, 562-580.

Lenhard, J.: 2004, 'Nanoscience and the Janus-Faced Character of Simulations', in: D. Baird, A. Nordmann & J. Schummer (eds.), *Discovering the Nanoscale*, Amsterdam: IOS Press, pp. 93-100.

Lenhard, J.: 2004, 'Nanoscience and the Janus-Faced Character of Simulations', in: D. Baird, A. Nordmann, J. Schummer (eds.), *Discovering the Nanoscale*, IOS Press, Amsterdam, pp. 93-100.

Lent, C.S., Tougaw, P.D., Porod, W., Bernstein, G.H., "Quantum Cellular Automata," *Nanotechnology*, Vol. 4, p. 49, 1993.

Lewak, S.E.: 2004, 'What's the Buzz? Tell Me What's A-Happening: Wonder, Nanotechnology, and Alice's Adventures in Wonderland', in: N.K. Hayles (ed.), *Nanoculture: Implications of the New Technoscience*, Bristol, UK: Intellect Books, pp. 201-.

Lide, D.R., ed. 1993-1994. *CRC Handbook of Chemistry and Physics*, 74th ed.

Lin-Easton, P.C.: 2001, 'It's Time for Environmentalists to Think Small ' Real Small: A Call for the Involvement of Environmental Lawyers in Developing Precautionary Policies Molecular Nanotechnology', *Georgetown International Law Review*, 14, 106-134.

López, J.: 2004, 'Bridging the Gaps: Science Fiction in Nanotechnology', *Hyle: International Journal for Philosophy of Chemistry*, 10(2), 129-152.

Lösch, A.: 2004, 'Nanomedicine and Space: Discursive Orders of Mediating Innovations', in: D. Baird, A. Nordmann & J. Schummer (eds.), *Discovering the Nanoscale*, Amsterdam: IOS Press, pp. 193-202.

Mackenzie, D.: 1996a, 'Economic and Sociological Explanations of Technological Change', in: *Knowing Machines: Essays on Technical Change*, MIT Press, Cambridge, MA, pp. 49-65.

Mackenzie, D.: 1996b, 'Marx and the Machine', in: *Knowing Machines: Essays on Technical Change*, MIT Press, Cambridge, MA, pp. 23-47.

Marshall, K.: 2004, 'Atomizing the Risk Technology', in: N.K. Hayles (ed.), *Nanoculture: Implications of the New Technoscience*, Bristol, UK: Intellect Books, pp. 147-160.

Martin, T.P., N. Malinowski, U. Zimmerman, U. Naher, and H. Schaber. 1993. *J. Chem. Phys.* 99:4210.

Martin, T.P., U. Naher, H. Schaber, U. Zimmerman. 1993. *Phys. Rev. Lett.* 70:3079.

Marty, A.: 2003, *Testimony before the Committee on Science on HR 766, Nanotechnology*

Research and Development Act, US Government Printing Office, Washington, DC.

Marx, K.: 1963 [1847], *The Poverty of Philosophy*, International Publishers, New York.

Mayer, S.: 2002, 'From genetic modification to nanotechnology: the dangers of 'sound science'', in: T. Gilland (ed.), *Science: Can We Trust the Experts?*, London: Hodder and Stoughton, pp. 1-15.

Mehta, M.D.: 2002, 'Nanoscience and Nanotechnology: Assessing the Nature of Innovation in These Fields', *Bulletin of Science, Technology & Society*, 22(4), 269-273.

Mehta, M.D.: 2004, 'From Biotechnology to Nanotechnology: What Can We Learn from Earlier Technologies?, *Bulletin of Science, Technology & Society*, 24, 34-39.

Meyer, M. & Kuusi, O.: 2004, 'Nanotechnology: Generalizations in an Interdisciplinary Field of Science and Technology', *Hyle: International Journal for Philosophy of Chemistry*, 10(2), 153-168.

Meyer, M.: 2000a, 'Does science push technology? Patents citing scientific literature', *Research Policy*, 29, 409-434.

Meyer, M.: 2000b, 'Patent citations in a novel field of technology: What can they tell about interactions of emerging communities of science and technology?', *Scientometrics*, 48, 151-178.

Meyer, M.: 2001a, 'Patent citations in a novel field of technology: An exploration of nano-science and nano-technology', *Scientometrics*, 51, 163-183.

Meyer, M.: 2001b, 'Socio-economic Research on Nanoscale Science and Technology: A European Overview and Illustration', in: M.C. Roco and W.S. Bainbridge (eds.), *Societal Implications of Nanoscience and Nanotechnology*, Dordrecht: Kluwer, pp. 217-241.

Meyer, M.; Persson, O.: 1998, 'Nanotechnology ' interdisciplinarity, patterns of collaboration and differences in application', *Scientometrics*, 47, 195'205.

Milburn, C.: 2002, 'Nanotechnology in the Age of Posthuman Engineering: Science Fiction as Science', *Configurations*, 10, 261-295.

Milburn, C.: 2004, 'Nano/Splatter: Disintegrating the Postbiological Body', *New Literary History*, 35 (in print).

Misa, T.: 1988, 'How Machines Make History and How Historians (and Others) Help Them to Do So', *Science, Technology, and Human Values*, 13, 308-331.

Mnyusiwalla, A.; Abdallah, S.D.; Singer, P.A.: 2003, 'Mind the gap: science and ethics in nanotechnology', *Nanotechnology*, 14, R9-R13.

Mody, C.C.M.: 2000, "A New Way of Flying': Différance, Rhetoric, and the Autogiro in Interwar Aviation', *Social Studies of Science*, 30, 513-543.

Mody, C.C.M.: 2001, 'A Little Dirt Never Hurt Anyone: Knowledge-Making and Contamination in Materials Science', *Social Studies of Science*, 31, 7-36.

Mody, C.C.M.: 2004, 'How Probe Microscopists Became Nanotechnologists', in: D. Baird, A. Nordmann & J. Schummer (eds.), *Discovering the Nanoscale*, Amsterdam: IOS Press, pp. 119-133.

Mody, C.C.M.: 2004, 'How Probe Microscopists Became Nanotechnologists', in: D. Baird, A. Nordmann, J. Schummer (eds.), *Discovering the Nanoscale*, IOS Press, Amsterdam, pp. 119-133.

Mody, C.C.M.: 2004, 'Instruments in Training: The Growth of American Probe Microscopy in the 1980s', in: D. Kaiser (ed.), *Pedagogy and the Practice of Science: Producing Physical Scientists, 1800-2000*, Cambridge, MA: MIT Press (forthcoming).

Mody, C.C.M.: 2004, 'Small, but Determined: Technological Determinism in Nanoscience', *Hyle: International Journal for Philosophy of Chemistry*, 10(2), 99-128.

Mody, C.C.M.: 2004, *Crafting the Tools of Knowledge: The Invention, Spread, and Commercialization of Probe Microscopy, 1960-2000*, Ph.D. dissertation, Cornell University.

Montemerlo, M.S., Love, J.C., Opiteck, G.J., Goldhaber, D. J., and Ellenbogen, J.C., "Technologies and Designs for Electronic Nanocomputers," MITRE Technical Report 96W0000044, The MITRE Corporation, McLean, VA, July 1996. For more information, send e-mail to nanotech@mitre.org.

Moor, J.H. & Weckert, J.: 2004, 'Nanoethics: Assessing the Nanoscale From an Ethical Point of View', in: D. Baird, A. Nordmann & J. Schummer (eds.), *Discovering the Nanoscale*, Amsterdam: IOS Press, pp. 301-310.

Munn Sanchez, E.: 2004, 'The Expert's Role in Nanoscience and Technology', in: D. Baird, A. Nordmann & J. Schummer (eds.), *Discovering the Nanoscale*, Amsterdam: IOS Press, pp. 257-266.

Nordmann, A.: 2003, 'Shaping the World Atom by Atom: Eine nanowissenschaftliche WeltBildanalyse', in: A. Grunwald (ed.), *Technikgestaltung zwischen Wunsch und Wirklichkeit*, Berlin: Springer, pp. 191-199.

Nordmann, A.: 2004, 'Molecular Disjunctions: Staking Claims at the Nanoscale', in: D. Baird, A. Nordmann & J. Schummer (eds.), *Discovering the Nanoscale*, Amsterdam: IOS Press, pp. 51-62.

Nordmann, A.: 2004, 'Molecular Disjunctions', in: D. Baird, A. Nordmann, J. Schummer (eds.), *Discovering the Nanoscale*, IOS Press, Amsterdam, pp. 51-62.

Nordmann, A.: 2004, 'Nanotechnology: Convergence and Integration', Presentation at *EuroNanoForum, Trieste, December 10, 2003* (Proceedings in preparation).

Nordmann, A.: 2004, 'Nanotechnology's WorldView: New Space for Old Cosmologies', *IEEE Technology and Society Magazine*, 23 (forthcoming).

Nordmann, A.: 2004, 'Social Imagination for Nanotechnology', in: European Commission (Community Health and Consumer Protection): *Nanotechnologies: A Preliminary Risk Analysis on the Basis of a Workshop, Brussels, 1-2 March 2004*, pp. 111-113 [www.europa.eu.int/comm/health/ph_risk/documents/ev_20040301_en.pdf].

Nordmann, A.: 2004, 'Was ist TechnoWissenschaft? ' Zum Wandel der Wissenschaftskultur am Beispiel von Nanoforschung und Bionik', in: T. Rossmann & C. Tropea (eds.), *Bionik ' Neue Forschungsergebnisse aus Natur-, Ingenieur- und Geisteswissenschaften*, Berlin: Springer, 2004 (forthcoming)

Nye, D.E.: 1994, *American Technological Sublime*, MIT Press, Cambridge, MA.

Nye, D.E.: 2003, *America as Second Creation: Technology and Narratives of New Beginnings*, MIT Press, Cambridge, MA.

Paschen, H.; Coenen, C.; Fleischer, T.; Grünwald, R.; Oertel, D. & Revermann, C.: 2004, *Nanotechnologie: Forschung, Entwicklung, Anwendung*, Berlin: Springer (*Nanotechnologie, TAB-Arbeitsbericht 92*, Berlin: Büro für Technikfolgen-Abschätzung beim Deutschen Bundestag).

Rademann, K., B. Kaiser, U. Even, F. Hensel. 1987. *Phys. Rev. Lett.* 59:2319.

Rao, C.N.R., B.C. Satishkumar, and A. Govindaraj. 1997. *Chem. Commun.* 1581.

Rao, M.B., and S. Sircar. 1993. *Gas Separation and Purification* 7:279.

Ratner, M.; Ratner, D.: 2003, *Nanotechnology: A Gentle Introduction to the Next Big Idea,* Prentice Hall, Upper Saddle River, NJ.

Reed, M.A., "Quantum Dots," *Scientific American,* January 1993, pp. 118-123.

Regis, E.: 1995, *Nano: The Emerging Science of Nanotechnology*, Little, Brown, Boston.

Reinhardt, C.: 2004, 'Chemistry in a Physical Mode: Molecular Spectroscopy and the Emergence of NMR', *Annals of Science*, 61, 1-32.

Rheinberger, H.-J.: 1997, *Toward a History of Epistemic Things: Synthesizing Proteins in the Test Tube*, Stanford University Press, Stanford.

Roberts, J.A.: 2004, 'Deciding the Future of Nanotechnologies: Legal Perspectives on Issues of Democracy and Technology', in: D. Baird, A. Nordmann & J. Schummer (eds.), *Discovering the Nanoscale*, Amsterdam: IOS Press, pp. 247-255.

Robinson, C.: 2004, 'Images in NanoScience/Technology', in: D. Baird, A. Nordmann & J. Schummer (eds.), *Discovering the Nanoscale*, Amsterdam: IOS Press, pp. 165-169.

Robison, W.L.: 2004, 'Nano-Ethics', in: D. Baird, A. Nordmann & J. Schummer (eds.), *Discovering the Nanoscale*, Amsterdam: IOS Press, pp. 285-300.

Roco, M.C. & Bainbridge, W.S. (eds.): 2001, *Societal implications of nanoscience and nanotechnology*, (Proceedings of a workshop organized by the National Science Foundation, September 28-29, 2000), Kluwer: Dordrecht [available online at http://itri.loyola.edu/nano/societalimpact/nanosi.pdf]

Roco, M.C.: 2003, 'Broader Societal Issues of Nanotechnology', *Journal of Nanoparticle Research*, 5, 181-189.

Roco, M.C.; Bainbridge, W.S. (eds.): 2002, *Converging Technologies for Improving Human Performance: Nanotechnology, Biotechnology, Information Technology and the Cognitive Science*, Arlington, VA: National Science Foundation.

Rohlfing, E.A., D.M. Cox, and A. Kaldor. 1984. *J. Chem. Phys*. 81:3846.

Rohrer, H.: 1995, 'The Nanometer Age – Challenge and Chance', *Microelectronic Engineering*, 27, 3-15.

Rosen, P.: 1993, 'The Social Construction of Mountain Bikes: Technology and Postmodernity in the Cycle Industry', *Social Studies of Science*, 23, 479-513.

Ruthven, D.M., S. Farooq, K.S. Knaebel. 1994. *Pressure swing adsorption*. New York: VCH Publishers.

Sarewitz, D.; Woodhouse, E.: 2003, 'Small is Powerful', in: A. Lightman, D. Sarewitz & Chr. Desser, (eds.), *Living with the Genie: Essays on Technology and the Quest for Human Mastery*, Washington, DC: Island Press, pp. 63-83.

Schiemann, G.: 2004, 'Dissolution of the Nature-Technology Dichotomy? Perspectives on Nanotechnology from an Everyday Understanding of Nature', in: D. Baird, A. Nordmann & J. Schummer (eds.), *Discovering the Nanoscale*, Amsterdam: IOS Press, pp. 209-213.

Schmidt, J.C.: 2004, 'Unbounded Technologies: Working Through Technological Reductionism of Nanotechnology', in: D. Baird, A. Nordmann & J. Schummer (eds.), *Discovering the Nanoscale*, Amsterdam: IOS Press, pp. 35-50.

Schuler, E.: 2004, 'Perception of Risks and Nanotechnology', in: D. Baird, A. Nordmann & J. Schummer (eds.), *Discovering the Nanoscale*, Amsterdam: IOS Press, pp. 279-284.

Schummer, J.: 2001, 'Ethics of Chemical Synthesis', *Hyle: International Journal for Philosophy of Chemistry*, 7, 103-124 [online: www.hyle.org/journal/issues/7/schummer.htm].

Schummer, J.: 2004, 'Interdisciplinary Issues in Nanoscale Research', in: D. Baird, A. Nordmann & J. Schummer (eds.), *Discovering the Nanoscale*, Amsterdam: IOS Press, pp. 9-20.

Schummer, J.: 2004, 'Interdisciplinary Issues in Nanoscale Research', in: D. Baird, A. Nordmann, J. Schummer (eds.), *Discovering the Nanoscale*, IOS Press, Amsterdam, pp. 9-20.

Schummer, J.: 2004, 'Multidisciplinarity, Interdisciplinarity, and Patterns of Research Collaboration in Nanoscience and Nanotechnology', *Scientometrics*, 59, 425-465.

Schummer, J.: 2004, 'Naturverhältnisse in der modernen Wirkstoff-Forschung', in: K. Kornwachs (ed.), *Technik ' System ' Verantwortung*, Münster: LIT, pp. 629-638.

Schummer, J.: 2004, 'Why do Chemists Perform Experiments?', in: D. Sobczynska, P. Zeidler, E. Zielonacka-Lis (eds.), *Chemistry in the Philosophical Melting Pot*, Peter Lang (Frankfurt/M.), 2004, pp. 395-410.

Schummer, J.: 2005, 'Reading Nano: The Public Interest in Nanotechnology as Reflected in Book Purchase Patterns", *Public Understanding of Science*, 14 (2) (forthcoming).

Schwarz, A.E.: 2004, 'Shrinking the 'Ecological Footprint' with NanoTechnoScience?', in: D. Baird, A. Nordmann & J. Schummer (eds.), *Discovering the Nanoscale*, Amsterdam: IOS Press, pp. 203-208.

Schwarz, R.B. 1998. Storage of hydrogen in powders with nanosized crystalline domains. In *R&D status and trends*, ed. Siegel et al.

Siegel, R.W., E. Hu, and M.C. Roco, eds. 1998. *R&D status and trends in nanoparticles, nanostructured materials, and nanodevices in the United States*. Baltimore: Loyola College, International Technology Research Institute. NTIS #PB98-117914.

Smith, M.R.; Marx, L. (eds.): 1994, *Does Technology Drive History? The Dilemma of Technological Determinism*, MIT Press, Cambridge, MA.

Soffer, A., J.E. Koresh, S. Saggy. 1987. United States Patent 4,685,940.

Stix, G.: 2001, 'Little Big Science: Nanotechnology Is All the Rage. But Will It Meet Its Ambitious Goals? And What The Heck Is It?', *Scientific American* 285, 32-37.

Stuckless, J.T, D.E. Starr, D.J. Bald, C.T. Campbell. 1997. *J. Chem. Phys.* 107:5547.

Suchman, M.: 2001, 'Envisioning Life on the Nano-Frontier', in: M.C. Roco and W.S. Bainbridge (eds.), *Societal Implications of Nanoscience and Nanotechnology*, Dordrecht: Kluwer, pp. 211-216.

Suchman, M.: 2002, 'Social Science and Nanotechnologies', M. Roco & R. Tomellini (eds.), *Nanotechnology 'Revolutionary Opportunities and Societal Implications*, Luxembourg: European Communities, pp. 95-99.

Sullivan, D.: 1994, 'Exclusionary Epideictic: *NOVA's* Narrative Excommunication of Fleischmann and Pons', *Science, Technology, and Human Values*, 19, 283-306.

Swami, N.: 2002, *Testimony before Senate Committee on Commerce, Science, and Transportation*, US Government Printing Office, Washington, DC.

Sweeney, A.E.; Seal, S. & Vaidyanathan, P.: 2003, 'The promises and perils of nanoscience and nanotechnology: Exploring emerging social and ethical issues', *Bulletin of Science, Technology & Society*, 23 (4), 236-245.

Tschope, A.S., W. Liu, M. Flyzani-Stephanopoulos, and J.Y. Ying 1995. *J. Catal*. 157:42.

Turner, F.: forthcoming, 'When the Counterculture Met Computer Networks and the New Economy: Revisiting the WELL and the Origins of Virtual Community', *Technology and Culture.*

Vermaas, P.E.: 2004, 'Nanoscale Technology: A Two-Sided Challenge for Interpretations of Quantum Mechanics', in: D. Baird, A. Nordmann & J. Schummer (eds.), *Discovering the Nanoscale*, Amsterdam: IOS Press, pp. 77-91.

Wald, C.A.: 2004, 'Working Boundaries on the NANO Exhibition', in: N.K. Hayles (ed.), *Nanoculture: Implications of the New Technoscience*, Bristol, UK: Intellect Books, pp. 83-108.

Weckert, J.: 2001, 'The Control of Scientific Research: The Case of Nanotechnology', *Australian Journal of Professional and Applied Ethics*, 3, 29-44.

Weil, V.: 2001, 'Ethical Issues in Nanotechnology', in: M.C. Roco and W.S. Bainbridge (eds.), *Societal Implications of Nanoscience and Nanotechnology*, Dordrecht: Kluwer, pp. 193-198.

Wenger, E.: 1998, *Communities of Practice: Learning, Meaning, and Identity*, Cambridge University Press, Cambridge, UK.

Index